Construction Economics
A new approach

Second Edition

Danny Myers

Taylor & Francis
Taylor & Francis Group

LONDON AND NEW YORK

First edition published 2004 by Spon Press

Second edition published 2008
by Taylor & Francis
2 Park Square, Milton Park, Abingdon, Oxon OX14 4RN

Simultaneously published in the USA and Canada
by Taylor & Francis
270 Madison Avenue, New York, NY 10016, USA

Reprinted 2010

Taylor & Francis is an imprint of the Taylor & Francis Group, an informa business

This work has been produced from typeset copy supplied by the Author
Printed and bound in Great Britain by The Cromwell Press Group, Trowbridge, Wiltshire

British Library Cataloguing in Publication Data
A catalogue record for this book is available from the British Library

Library of Congress Cataloguing in Publication Data
Myers, Danny.
        Construction economics: a new approach / Danny Myers. – 2nd ed.
        p.cm.
        Includes bibliographical references and index.
        1. Construction industry. 2. Construction industry–Management I. Title.
HD9715.A2M94 2009
338.4'7624–dc22
2008033654

ISBN10: 0-415-46228-2 (hbk)
ISBN10: 0-415-46229-0 (pbk)

ISBN13: 978-0-415-46228-0 (hbk)
ISBN13: 978-0-415-46229-7 (pbk)

Contents

LIST OF TABLES AND FIGURES

Tables

Figures

ACKNOWLEDGEMENTS

Although the book cover implies this is all my own work, it was not achieved alone. In chronological order, the staff and students that I have worked with over the years have made their mark. In particular, Melanie Dunster and Kevin Burnside, economists with a keen eye for language and detail, helped during the planning phase.

Next, to bring life to a manuscript is no easy task and this would have been impossible without the graphic skills, and patience of Chris Wade, who designed the text and artwork; his persistent attention to detail is evident in the following pages.

A major concern from the outset was to create a text that was easy to read and use. This is some challenge in the subject area, but perceptive and detailed editing by Paul Stirner helped to move the project towards this goal. His comments, questions and insights add to the clarity and rigour of the text, and have certainly helped to make the completed manuscript as accessible as possible.

Finally I want to acknowledge the support of the publishers and printers; to name just two from the team that I have contacted at Taylor Francis, Katy Low and Faith McDonald patiently co-ordinated, managed and provided reassurance from commissioning the new edition through to publication.

I hope you find the finished product interesting and relatively easy to use. If any errors or omissions remain, I apologise for these in advance, and would be grateful for correspondence bringing them to my attention. Enjoy the book!

Danny Myers
August, 2008

1 An Introduction to the Basic Concepts

This book is written for students from many backgrounds: architecture, surveying, civil engineering, mechanical engineering, structural engineering; construction, project or estate management, property development, conservation and, even, economics. Economics students may find it possible to skip over some of the standard analysis, but should be forewarned that in many ways construction is quite distinct from other sectors of the economy. An important aim of this text is to draw out these distinctions and clarify the unique nature of the industry. In this first chapter we begin to outline the main characteristics of firms involved in construction markets, introducing the complexity of the construction process and diversity of activities. As the chapter develops you will sense that there are a number of possible ways to describe the construction industry. Table 1.1 identifies a range of activities that can be included in a broad definition of the industry. By contrast, Table 1.2 (see page 11) divides the construction process into a number of professional stages and Table 1.3 (see page 19) outlines a simple classification system that narrowly defines the industry as firms that just construct and maintain buildings and infrastructure.

Table 1.1 The construction industry – broadly defined

	The key actors include:
✓	Suppliers of basic materials, e.g. cement and bricks
✓	Machinery manufacturers who provide equipment used on site, such as cranes and bulldozers
✓	Manufacturers of building components, e.g.windows and doors
✓	Site operatives who bring together components and materials
✓	Project managers and surveyors who co-ordinate the overall assembly
✓	Developers and architects who initiate and design new projects
✓	Facility managers who manage and maintain property
✓	Providers of complementary goods and services such as transportation, distribution, demolition, disposal and clean-up

Source: Adapted from Manseau and Seaden (2001: 3–4)

The aim of the text is to demonstrate that underlying the construction process, from conception to demolition, is a lot of useful economics. Economics should not be regarded as a discipline solely related to the appraisal of costs. The subject matter is far broader, and this text introduces a number of branches of economic theory. These have been selected to provide fresh insights into the performance of construction firms and a greater understanding of the need for a more holistic approach if the industry is to contribute to an efficient and sustainable economy in the future. These economic ideas should inform the work of all professionals concerned with the construction and maintenance of buildings and infrastructure – and, in particular, the way that they think.

The next section explains some of the key concepts used by economists. Further clarification is provided in the glossary at the back of the book, where all the economic terms highlighted in the text and other concepts and ideas relevant to construction economics are defined.

INTRODUCING CONSTRUCTION ECONOMICS

Construction economics – like pure economics, its mainstream equivalent – is concerned with the allocation of scarce resources. This is far more complex than it at first appears. Many of the world's resources (factors of production such as land, labour, capital and enterprise) are finite, yet people have infinite wants. We are, therefore, faced with a two-pronged problem: at any point in time there is a fixed stock of resources, set against many wants. This problem is formally referred to as **scarcity**. In an attempt to reconcile this problem, economists argue that people must make careful choices – choices about what is made, how it is made and for whom it is made; or in terms of construction, choices about what investments are made, how these are constructed and on whose behalf. Indeed, at its very simplest level, **economics** is 'the science of choice'.

When a choice is made, therefore, some other thing that is also desired has to be forgone. In other words, in a world of scarcity, for every want that is satisfied, some other want, or wants, remain unsatisfied. Choosing one thing inevitably requires giving up something else. An opportunity has been missed or forgone. To highlight this dilemma, economists refer to the concept of **opportunity cost**. One definition of opportunity cost is:

> the value of the alternative forgone by choosing a particular activity.

Once you have grasped this basic economic concept, you will begin to understand how economists think – how they think about children allocating their time between different games; governments determining what their budgets will be spent on; and construction firms deciding which projects to proceed with. In short, opportunity costs enable relative values to be placed on all employed resources.

This way of thinking emphasises that whenever an economic decision is made there is a **trade-off** between the use of one resource for one or more alternative uses. From an economic viewpoint the value of a trade-off is the 'real cost' – or opportunity cost – of the decision. This can be demonstrated by examining the opportunity cost of reading this book. Let us assume that you have a maximum of four hours each week to spend studying just two topics – construction economics and construction technology. The more you study construction economics, the higher will be your expected grade; the more you study construction technology, the higher will be your expected grade in that subject. There is a trade-off, between spending one more hour reading this book and spending that hour studying technology. In this example there is fixed trade-off ratio. In practice, however, some people are better suited to some subjects than others and the same thing can be applied to resources. As a general rule, therefore, resources are rarely equally adaptable to alternative projects.

In construction, or any other economic sector, it is rare to experience a constant opportunity-cost ratio, in which each unit of production can be directly adapted to an alternative use. It is far more usual in business trade-off decisions to see each additional unit of production cost more in forgone alternatives than the previously produced unit. This rule is formally referred to as the **law of increasing opportunity costs**. This can be illustrated with the 'guns or butter' argument – this states that, at any point in time, a nation can have either more military goods (guns) or civilian goods (butter) – but not in equal proportions. For example, consider the hypothetical position in which all resources in the first instance are devoted to making civilian goods, and the production of military goods is zero. If we begin production of military goods, at first production will increase relatively quickly, as we might find some engineers who could easily produce military goods and their productivity might be roughly the same in either sector. Eventually, however, as we run out of talent, it may become necessary to transfer manual agricultural labour used to harvesting potatoes to produce military goods – and their talents will be relatively ill-suited to these new tasks. We may find it necessary to use fifty manual labourers to obtain the same increment in military goods output that we achieved when we hired one sophisticated engineer for the first units of military goods. Thus the opportunity cost of an additional unit of military goods will be higher when we use resources that are inappropriate to the task. By using poorly suited resources, the cost increases as we attempt to produce more and more military goods and fewer and fewer civilian goods.

The law of increasing opportunity costs is easier to explain using a **production possibility curve**. Using these curves, it is possible to show the maximum amount of output that can be produced from a fixed amount of resources. In Figure 1.1 (see page 4) we show a hypothetical trade-off between units of military goods and civilian goods produced per year. If no civilian goods are produced, all resources would be used in the production of military goods and, at the other extreme, if no military goods are produced, all resources would be used to produce civilian goods. Points A and F in Figure 1.1 represent these two extreme positions. Points B, C, D and E represent various other combinations that are possible. If these points are connected with a smooth curve, society's production possibilities curve is obtained, and it demonstrates the trade-off between the production of military and civilian goods. These trade-offs occur on the production possibility curve. The curve is bowed outwards to reflect the law of increasing opportunity cost. If the trade-off is equal, unit for unit, the curve would not bow out, it would simply be a straight line. Other interesting observations arising from the production possibility curve are shown by points G and H. Point G lies outside the production possibility curve and is unattainable at the present point in time, but it does represent a target for the future. Point H, on the other hand, lies inside the production possibility curve and is, therefore, achievable, but it represents an inefficient use of available resources.

Figure 1.1 The trade-off between military goods and civilian goods

Points A to F represent the various combinations of military and civilian goods that can be achieved. Connecting the points with a smooth line creates the production possibility curve. Point G lies outside the production possibility curve and is unattainable at the present time; point H represents an inefficient use of resources at the present time.

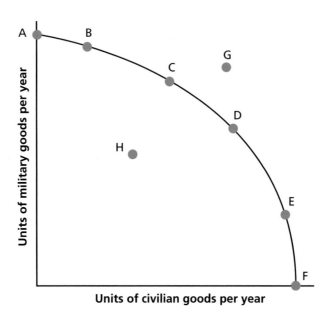

There are a number of assumptions underlying the production possibility curve. The first relates to the fact that we are referring to the output possible on a *yearly* basis. In other words, we have specified a time period during which production takes place. Second, we are assuming that resources are fixed throughout this time period. To understand fully what is meant by a fixed amount of resources, consider the two lists that follow, showing (a) factors that influence labour hours available for work and (b) factors that influence productivity, or the output per unit of input.

FACTORS INFLUENCING LABOUR HOURS AVAILABLE FOR WORK

The number of labour hours available for work depends on the nature of human resources in society. This is determined by three factors:

- the number of economically active people that make up the labour force – this depends on the size of the population and its age structure, as children and retired persons will be economically inactive
- the percentage of the labour force who then choose to work
- prevailing customs and traditions (such as typical length of the working week, number of bank holidays, etc.).

FACTORS INFLUENCING PRODUCTIVITY

There are a number of factors influencing the productivity of an economy or sector of the economy:

- the quantity and quality of natural and man-made resources
- the quality and extent of the education and training of the labour force
- the levels of expectation, motivation and wellbeing
- the commitment to research and development.

The third and final assumption that is made when we draw the production possibility curve is that efficient use is being made of all available resources. In other words, society cannot for the moment be more productive with the present quantity and quality of its resources. (The concept of efficiency is examined more closely in Chapters 2, 5, 6, 7 and 8.)

According to several government reports (Egan 1998 and 2002; NAO 2001, 2005 and 2007), given the existing level of resources in construction it should be possible to increase productivity by at least 10 per cent. In other words, a production possibility curve representing all construction activities could be pushed out to the right, as shown in Figure 1.2 (see page 6). Several common sets of problem are identified as the root cause of this inefficiency. First, the industry demonstrates a poor safety record and an inability to recruit good staff. Second, there appears to be no real culture of learning from previous projects, and no organised career structure to develop supervisory and management grades. Third, concern is expressed about the poor level of investment into research and development that restricts the industry's ability to innovate and learn from best practice. The fourth, and possibly most worrying, problem is the fact that technology (in the sense of IT, prefabrication and off-site assembly) is not used widely enough across the construction sector.

Another plausible scenario suggested by the production possibility curve approach is that the construction industry may at present be working within the boundary of its production curve (say, point A in Figure 1.2). In which case, an increase in output could be simply achieved by greater efficiency. Supply constraints need to be reduced, the problems identified by the government reports resolved, and the factors generally acknowledged to increase productivity (listed above) must be addressed to achieve the full potential of the industry. Both these scenarios are shown in Figure 1.2 and they support the idea that the level of productivity in the construction industry needs to improve.

In very general terms, therefore, the study of economics (and construction economics) is concerned with making efficient use of limited resources to maximise output and satisfy the greatest possible number of wants. In short, the basis of the subject rotates around the concepts of choice, scarcity and opportunity cost.

Figure 1.2 Increasing output and the production possibility curve

In this diagram we show two scenarios: (a) improved productivity shifts the entire production possibility curve outwards over time; (b) output can be achieved more efficiently by moving to a position of full potential on the actual production possibility curve.

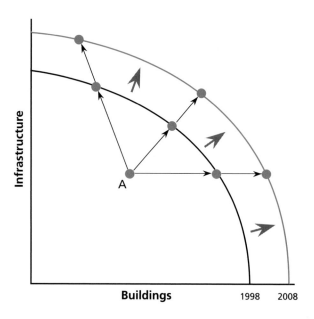

In modern society, economics is involved in all activities leading to the production of goods and services. Consequently a range of specialisms have evolved out of mainstream economics, such as transport economics, health economics, business economics, financial economics, agricultural economics, labour economics, international economics and, even, ecological economics. Hence it is not particularly surprising that many students in the twenty-first century are expected to read something called construction economics as part of their degree course. What is surprising, however, is that other vocationally oriented degrees do not have a similarly developed economics specialism. For example, students reading for degrees in catering, sports and leisure, publishing, retailing or computing do not benefit from a range of specialised literature in economics.

The reasons usually stated for construction warranting its own specialised economics is accounted for by the sheer size of the industry, its profound contribution to a nation's standard of living and its products' unique characteristics. Put very simply, the industry has five distinct qualities.

- The physical nature of the product is large, heavy and expensive.
- The construction industry is dominated by a large number of relatively small firms, spread over a vast geographical area.
- Demand for activity within the industry is directly determined by the general state of the economy as a whole.

- The method of price determination is unusually complex due to the tendering process used at various stages.
- Most projects can be considered as a 'one-off', as there is usually some defining quality that make them in some ways unique.

These qualities alone have justified a number of dedicated academic publications. In 1974 the first edition of Patricia Hillebrandt's *Economic Theory and the Construction Industry* was published. Subsequently several other titles have appeared – for details see the reference section at the back of this book – in particular the two-part text co-authored by Graham Ive and Stephen Gruneberg (2000). In 1982, *Construction Management and Economics* a specialist refereed journal began to report on research contributing to the new subject specialism. This journal is published monthly and more than 1200 papers have appeared; several of them are drawn upon to support this text. Some extracts have been selected from these papers as case study readings to consolidate the three sections of this textbook. Another relevant academic journal is *Building Research and Information*. This has an interdisciplinary focus, with linkages made between the built, natural, social and economic environments. Consequently many of the papers in this journal contribute to our understanding of how buildings and infrastructure impact on ecology, resources, climate change and sustainable development; appropriately several examples are cited as references.

Alongside these academic developments, there has also been a succession of government reports investigating the problems of the construction industry (for example, see Latham 1994; Egan 1998; DETR 2000; National Audit Office 2001; Fairclough 2002; HM Government 2008). These reports have highlighted the inefficiency caused by the sheer scale and complexity of the construction industry. A recurring recommendation is the need for the construction process to be viewed in a holistic way by a multidisciplinary team. This reflects the fact that construction draws knowledge from many areas, and an important but undervalued area is economics. Indeed, it is commonly observed that far too many projects run over budget and are delivered late, with a general disrespect for the client. Clearly it should not be acceptable for construction projects to fail cost-wise, time-wise or client-wise. An authoritative study by Professor Flyvbjerg (2003: 16–26) of 258 major public transport infrastructure projects constructed across Europe, USA, Japan and developing countries between 1927 and 1998 suggests that on average costs overrun by approximately 30 per cent, deadlines are missed by as much as ten years and the expected level of demand fails to meet targets by around 40 per cent.

Each of the construction economics texts that have been published to date conveys a slightly different emphasis. For example, Hillebrandt (1974, 2000) defines construction economics as the application of the techniques and expertise of economics to the study of the construction firm, the construction process and the construction industry. Whereas the preference of Ive and Gruneberg (2000: xxiii) is for a slightly less orthodox approach, adapting traditional economic models to capture local circumstances even if that means losing the ability to generalise about the economy at large. As a result, there is no coherent conceptual consensus about what constitutes the precise nature of construction economics. As George Ofori (1994: 304) bluntly concluded in his seminal review of the subject: 'Construction economics cannot be regarded as a bona-fide academic discipline. It lacks a clear indication of its main concerns and content.'

The purpose of this text is to address this lack of consensus and make the case for a coherent economic vocabulary. The crux of the argument for this new approach is the increasing importance of strategies aimed at achieving **sustainable construction**. In other words, there is an increasing recognition that the industry makes an important contribution to a country's economic, social and environmental wellbeing.

INTRODUCING SUSTAINABLE CONSTRUCTION

The UK government published its first strategy for more sustainable construction, *Building a Better Quality of Life,* in April 2000. This document aimed to provide a catalyst for change in the approach to construction processes. Subsequently it has been revised and extended, see for example the *Strategy for Sustainable Construction* published in June 2008. Similar agendas have emerged in Europe, North America and some developing countries (see Chapter 15 for examples and references). Sustainable construction can be described in simple terms as comprising:

- efficient use of resources
- effective protection of the environment
- economic growth
- social progress that meets the needs of everyone.

Each of these strands is underpinned by economic concepts, which provide the rationale for this book.

Part A   Effective use of resources
This deals with microeconomics, and outlines the various ways of efficiently allocating resources between competing ends. In this section the prime focus is concerned with the determinants of demand and supply for infrastructure, housing, industrial buildings, commercial property, and repair and maintenance.

Part B   Protection and enhancement of the environment
This section considers failures of the market system, drawing upon various environmental economic concepts and tools to encourage future members of the construction industry to evaluate projects by more than just financial criteria.

Part C   Economic growth that meets the needs of everyone
This section incorporates coverage of the broader macroeconomic scene. It outlines the various government objectives that need to be achieved alongside sustainable construction. It highlights the difficulty of managing an economy and the need for professionals working in the construction industry to acquire an economic vocabulary.

Key Points 1.1

❑ The construction industry can be described in a number of ways – for example, review the broad range of activities listed in Table 1.1 (page 1).

❑ Construction has five distinguishing characteristics: (a) each project is regarded as a unique one-off product; (b) the industry is dominated by a large number of relatively small firms; (c) the general state of the economy influences demand; (d) prices are determined by tendering; and (e) projects are characterised by their 'lumpiness' in terms of their scale and expense.

❑ The basis of economics rotates around the concepts of choice, scarcity and opportunity cost. Hence, economics is the study of how we make choices.

❑ Any use of a resource involves an opportunity cost because an alternative use is sacrificed.

❑ The graphic representation of the trade-offs that must be made can be displayed in a production possibility curve.

❑ Sustainable construction is a strategy aimed to encourage the industry to (a) use resources more efficiently, (b) limit the environmental impact of its activities, and (c) produce buildings and infrastructure that benefit everyone.

INTRODUCING ECONOMIC VOCABULARY

The discipline of economics employs its own particular methodology and language. Consequently for the complete beginner it is necessary at the outset to clarify a few meanings.

Resources

Resources can be defined as the inputs used in the production of those things that we desire. Economists tend to refer to these resources as **factors of production** to highlight the fact that only by combining various factors can goods and services be produced. The factors of production are usually categorised into three general groups; namely, land, capital and labour – and sometimes the entrepreneur is specifically identified as a fourth entity. The point is that quantities of each factor are needed to make any good or service. To construct buildings or infrastructure, for example, labour is required to develop a plot of land, and plant and equipment, which may be hired or bought, is required to facilitate the process. To put it another way, land and labour are always combined with manufactured resources in order to produce the things that we desire. The manufactured resources are called **capital**, or more precisely physical capital, and consist of machines and tools.

The contribution of labour to the production process can be increased. Whenever potential labourers undergo training and learn new skills, their contribution to productive output will increase. When there is this improvement in human resources, we say that **human capital** has been improved. A relevant example is the effect that good trained management can have on the efficiency of a whole project. Indeed, according to Hillebrandt (2000: 104) management expertise is one of the scarcest resources of the construction industry throughout the world.

With each new project there is a choice to be made about the materials that will be used and the proportion of labour, plant and equipment required. In most instances, construction tends to be dominated by input costs relating to materials, components and labour. The importance, however, of the entrepreneur should not be overlooked, as without a dedicated resource managing and co-ordinating the other factors of production, virtually no business organisation could operate. In other words, an entrepreneur is sometimes regarded as a special type of human resource associated with the ability to make business decisions, take risks and foster innovation. In a small construction firm the manager-proprietor would be the entrepreneur; in a joint stock company the shareholders would take on that responsibility.

Each factor of production can be regarded as receiving a specific form of income. A landlord providing the use of land receives rent. Owners of physical (and monetary) capital are rewarded, directly or indirectly, and earn some form of interest payment to cover credit arrangements. Workers receive wages (salaries), and entrepreneurs gain profit. The distribution of these factor rewards (factor incomes) formed an important point of focus for the classical economists. Ricardo's work (1817) suggested that the rewards paid to the agricultural landlord determined all other payments and this inevitably led to a tension between the interests of the landlord and those of the consumer and manufacturer. Equally Marx (1844) was concerned about the inequalities that were rewarded to labour as he claimed that they were exploited by the owners of capital and land, as he observed: a worker cannot supplement his income with ground rent or interest on capital. For general introductory purposes, however, the significance of dividing income payments into four sets of factor rewards will become evident when we consider the measurement of **national income** in Chapter 13.

Market Systems

The concept of the **market** is rather abstract in the sense that it encompasses the exchange arrangements of both buyers and sellers for a particular good or service. Consequently, we can envisage many markets for specific building materials, housing, professional services, etc. The recurrent feature of any market is the exchange of information about factors such as price, quality and quantity. The difference, however, between one market and the next is the degree of formality in which it functions. The stock market in any Western economy, for example, provides instant information worldwide about the prices and quantities of shares being bought and sold during the current trading period. By contrast, construction markets are less structured and more informal, and they are usually determined by geographical location.

The construction industry is concerned with producing and maintaining a wide variety of durable buildings and structures, and as a consequence, there are many construction markets. As Drew and Skitmore (1997: 470) concluded in their analysis of the competitive markets for construction: 'The construction industry is highly fragmented, with the dominant firm being the small contractor.' The type of construction – particularly in terms of its size and complexity, its geographical location, and the nature of the client – will define the market in each case.

Let us consider in a little more detail what traditionally happens when a new project begins. Usually a contractor undertakes to organise, move and assemble the various inputs, and as such provides a service – a service of preparing the site before work commences, and assembling and managing the process thereafter. Subsequently, various subcontractors add their services – such as plumbing, painting, plastering, glazing, roofing, or whatever the specific job requires. As a result the typical project process can easily become a series of 'separate' operations undertaken by various parties as set out in Table 1.2.

Table 1.2 Parties traditionally supplying a construction project

Parties Involved in Supply	Responsibilities
Architects and Designers	Provide specialist advice concerning structural, electrical, mechanical and landscape details. Identify key specifications.
Project Manager	Manages project in detail. Liaises between the client and the construction team.
Cost Consultant	Prepares bills of quantities, cost plans, etc.
Main Contractor	Manages work on site.
Subcontractors	Supply specialist skills.
Suppliers	Provide building materials and related components.

The level of competition for all this work depends upon the complexity of the construction (which to some extent will be reflected in the cost per square metre). The idea of complexity is particularly important in construction markets as it determines the number of businesses interested in competing for the work. In most cases, firms will not bid for work beyond their local district as the costs of transporting materials, plant and labour is relatively expensive. Travelling is unnecessary when the same type of work is available in the firm's own neighbourhood or catchment area. If, however, the construction project is very complex and/or very large, the costs per square metre are likely to increase and the relative costs of transport in relation to the total costs will decrease. The market catchment area for this highly specialised work will broaden. The following formula may make this clearer:

complexity + large size = competing firms from a wider geographical area

The converse of this rule explains why construction markets are so often dominated by small local firms subcontracting for work in or near their home towns. Indeed, it is only the biggest firms that can manage to compete on a national or international basis. Markets in the construction industry should, therefore, be defined as comprising those firms that are willing and able to compete for a contract in a specific geographical area. In other words, the total number of firms interested in work of a particular type can be referred to as comprising the local market.

In construction the services of one firm are often easy to substitute by contracting another firm with the same type of expertise. To the extent that prices in construction markets often find their own level, the theory behind this behaviour is examined in Chapters 3 to 8. For the moment, it will suffice to understand that the

Figure 1.3 A complex set of markets for one building project

In the flow diagram the building project is represented as a sequence of stages. Each stage is completed by a number of firms supplying their services.

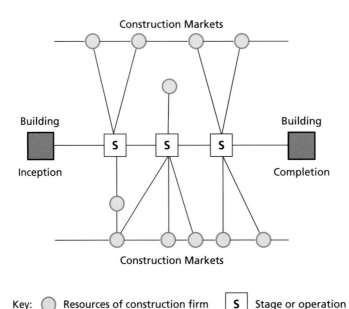

Source: Adapted from Turin (1975: 70)

market for construction refers to a diverse and broad range of activities made up of many markets. To emphasise this point, consider the flow diagram set out in Figure 1.3 (page 12).

THE EXAMPLE OF ONE BUILDING PROJECT

In Figure 1.3 we represent a set of markets that could be involved in the construction of a small commercial building. The construction process from inception to completion is shown to comprise a number of separate markets. The completion of each operation, or stage, is the concern of several construction firms competing to supply materials, components, labour, etc. Figure 1.3 highlights the number of fragmented activities involved in completing just one small building. Each independent firm, in effect, is more concerned with its specific contribution than the project as a whole. In many ways, the next project in a firm's market sector may well be competing for its attention while it is still finishing the present project.

We discuss the characteristics of fragmentation and the resulting poor flow of information in Chapters 6 and 9 respectively. We also identify these issues as a problem to resolve in Chapter 15, where we analyse loosely connected activities as a barrier to achieving sustainable construction.

METHODOLOGY

This introductory chapter aims to explain what construction economics is all about. Therefore, apart from identifying the central concepts, we need to consider the methods employed by economists, as the approach taken to a discipline also helps to specify the nature of the subject. In general terms, economics is a social science and it attempts to make use of the same kinds of methods as other sciences, such as biology, physics and chemistry. Like these other sciences, economics uses models or theories.

> Economic models are simplified representations of the real world that we use to understand, explain and predict economic phenomena.

These models may take on various forms such as verbal statements, numerical tables and graphs – and, at the more advanced level, mathematical equations. For the most part, the models presented in this text consist of verbal statements and graphs.

A particular challenge faced by students of construction economics is that many of the processes in the industry do not lend themselves easily to generalisations and models. First, the construction industry involves a large variety of interests and parties that makes the process rather complex and plagued with unwarranted assumptions about what is possible. Second, economic analysis is only one of the disciplines contributing to the process as a whole. And, third, there is a distinct lack of vision about the role of construction in society and how it could better serve its clients. As Professor Duccio Turin poetically observed:

> [T]he building process is a world of 'as if'. It is 'as if' the client knew what he wanted when he commissioned the building from a designer; it is 'as if' the designer was in a position to advise the client on the best value-for-

money he could obtain in the market; it is 'as if' contractual procedures were devised to ensure that the client would get the best possible deal from the profession and from the market place; it is 'as if' the manufacturer of building materials and components knew in advance what is expected of him and geared his production to such expectation; it is 'as if' the contractor knew how his resources were used, was in a position to control them, and was able to use this experience on his next job.

(Turin 1975: xi).

Although this summary of the industry was expressed nearly 30 years ago, as the text unfolds you will realise some striking similarity between then and now. It is this complex, fragmented and conservative nature that gives the subject matter of construction economics its appeal – as economists seek to unravel these seemingly unconnected threads of random behaviour. Economic models seek to identify the interrelationship between the key variables and simplify what is happening in the sector. So although some economic models may at first appear abstract, they do have practical applications. The important point we are trying to clarify is that an economic model cannot be criticised as unrealistic merely because it does not represent every last detail of the real world that it is seeking to analyse. If the model elucidates the central issues being studied, then it is worthwhile. For example, students may be expected to commence their course by completing an assignment based on a theoretical economic model of competition in the marketplace. This provides a simple introduction to the economic framework and the opportunity to demonstrate how construction deviates from or reflects this reference point. In short, the model provides a starting point – it enables us to proceed.

Following the recommendations of the Fairclough report (2002: 34) the construction industry should favour models that prioritise strategies aimed to improve sustainability, competitiveness, productivity and value to clients. In Part A, we present models of market behaviour that encourage a far better grasp of the meaning and purpose of efficiency, competition and profit. In Part C, we introduce a model of aggregation to study the operation of the whole economy that brings a fresh dimension to productivity by reviewing the total output of construction and reflecting on its contribution to the total output of an economy. In Part B, we bring the environment into the traditional model as a key variable for construction and the economy to consider. When the whole book has been studied, we identify a significant number of concepts that underpin an understanding of sustainability.

This leaves the precise nature and details of the models to emerge as the book unfolds and their purpose should become self-evident. Once we have determined that a model does predict real-world phenomena, then the scientific approach requires that we consider *evidence* to test the usefulness of a model. This is why economics is referred to as an empirical science – empirical meaning that real data is gathered to confirm that our assumptions are right.

An Example of an Economic Model

Before closing this section on models, we review one specific example to analyse and explain how income flows around an economy. Economists begin their explanation by ignoring the government sector, the financial sector and the overseas sector – that is, the **circular flow model** represents a simplified, scaled-down economy in which relationships are assumed to exist only between households and businesses.

To make the model effective, it is assumed that households sell factors of production to businesses and in return receive income in the form of wages, interest, rents and profits. This is shown in the bottom loop in Figure 1.4. The businesses sell finished goods and services to households in exchange for household expenditure. This is shown in the top loop in Figure 1.4. These assumptions are reasonably realistic. Businesses will only make what they can sell. Production will necessitate buying in land, labour, capital and enterprise, and the monies paid for these factors of production will generate respective income payments.

Already, without building in any of the complications of the real world, we begin to sense several insights or starting points. Clearly there is a close relationship between the income of a nation, its output and the level of expenditure, and we shall investigate this further in Chapter 13. Also, we can see how money enables households to 'vote' for the goods and services desired, and this will be developed further in Chapters 2 to 5. The perceptive reader will note that the model fails to

Figure 1.4 The circular flow model: a two sector economy

In this simplified model there are only households and businesses. Goods and services flow in one direction in return for money. This exchange can be thought of as a circular flow.

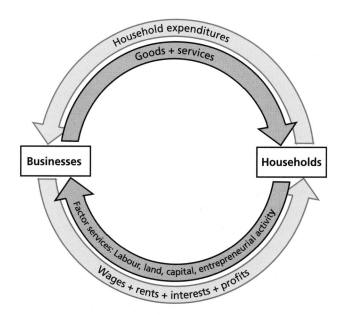

include any reference to the environment and, as we explained above, this represents the contents of Part B. The model also explicitly excludes reference to the role of governments, overseas economies and financial institutions and these aspects are included in Part C. Again we can see how the model enables us to progress into the subject.

THE ROLE OF GOVERNMENTS

The importance of the construction industry to the overall wellbeing of the economy means that most governments are concerned that it becomes a highly efficient sector of the economy. As a consequence, the government's role as a client, regulator, policy-maker and a sponsor of change is raised at several points throughout the text. In particular, it is discussed in Chapter 2 where we consider the power of the government in allocating resources, throughout Part C where we focus on macroeconomics and in Chapter 15 where we review the possibility of achieving the government agenda of sustainable development.

ENVIRONMENTAL ECONOMICS

In Part B we look at effective protection of the environment. This forms a key part of any text concerning sustainability. **Environmental economics** is important for several reasons: first, because the environment has an intrinsic value that must not be overlooked; second, because the sustainability agenda extends the time horizon of any analysis to make sure that decision-makers today consider some form of equity with future generations; and, third, demands must be viewed on a whole-life basis and this is particularly important in the context of products that last for more than 30 years. Any model of analysis that seeks to identify general principles of sustainable construction must include, at the very least, these three dimensions. In Chapters 9 to 11 we explore these issues and related concepts.

Microeconomics and Macroeconomics

Economics is typically divided into two types of analysis: **microeconomics** and **macroeconomics**. Consider the definitions of the two terms.

> Microeconomics is the study of individual decision-making by both individuals and firms.

> Macroeconomics is the study of economy-wide phenomena resulting from group decision-making in entire markets. As such, it deals with the economy as a whole.

One way to understand the distinction between these two approaches is to consider some generalised examples. Microeconomics is concerned with determining how prices emerge and change, and how firms respond. It involves the examination of the effects of new taxes, the determination of a firm's profit-maximising level of production, and so on. In other words, it concerns the economic behaviour of individuals – such as clients, contractors, surveyors and engineers – in various markets. We study this type of analysis in Part A. In contrast, questions relating to the rate of inflation, the amount of national unemployment, the growth rate of the whole economy and numerous other economy-wide subjects all fall in the realm of macroeconomic analysis. In other words, macroeconomics deals with aggregates or totals, and this forms the basis of the three chapters that comprise Part C.

You should be aware, however, of the blending together of microeconomics and macroeconomics in modern economic theory. Modern economists are increasingly using microeconomic analysis – the study of decision-making by individuals and by firms – as the basis for macroeconomic analysis. They do this because, even though aggregates are being examined in macroeconomic analysis, those aggregates are made up of the actions of individuals and firms. The study of any specific industry involves both microeconomic and macroeconomic approaches; particularly when the industry is multi-product, and has national and international significance.

Throughout this text the interaction between the construction sector and the other sectors of the economy is a constant reference point. In some texts a sectoral approach is referred to as **mesoeconomics**, derived from the Greek word *mesos* meaning intermediate. This is done to make it clear that the study of any specific sector or industry inevitably falls between the conventional microeconomic and macroeconomic categories. (These two terms are also of Greek derivation: *macros* – large and *micros* – small.)

Consequently, to gain a comprehensive understanding of construction activity, it is advisable to embrace three perspectives – a broad macro overview of the economy, a specific sectoral study of the industry, and a detailed microanalysis of the individual markets in which construction firms operate. Studying the complete text, therefore, should provide a greater understanding of winning and completing projects in an efficient and sustainable manner. Figure 1.5 summarises how these three elements contribute to a fuller understanding of a project. In many ways, it models the overall approach of the text.

Figure 1.5 A model for construction economics: a new approach

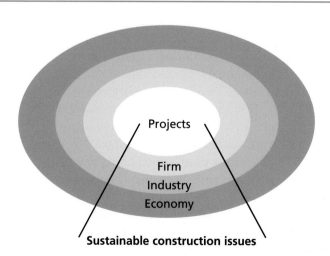

Key Points 1.2

❑ We need to use scarce resources, such as land, labour, capital and entrepreneurship, to produce any economic good or service.

❑ The exchange of information between buyers and sellers about factors such as price, quality and quantity happens in a market. Construction is made up of a diverse range of markets, as the industry comprises a large number of relatively small firms.

❑ Every economic model, or theory, is based on a set of assumptions. How realistic these assumptions are is not as important as how effective they make the model or theory.

❑ Microeconomics involves the study of individual decision-making. Macroeconomics involves the study of aggregates. Mesoeconomics combines the territory shared by microeconomics and macroeconomics to study a specific sector such as construction.

INTRODUCING CONSTRUCTION INDUSTRY ACTIVITY

The system of industrial classification used for statistical and government purposes favours a narrow definition of the construction industry that includes only firms that are involved with building and civil engineering. This categorisation is derived from the United Nations International Standard of Industrial Classification (ISIC). There are also American and European equivalents: the North American Industry Classification System (NAICS) and the General Industrial Classification of Economic Activities – otherwise known as NACE. In other words, firms generally recognised as officially comprising the construction sector tend to embrace a range of 'on-site' activities including those relating to infrastructure, new construction, repair, maintenance and (eventually) demolition. Table 1.3 shows the type of work that is classified into these various sectors and Table 1.4 gives some indication of the monetary value of these different activities in Great Britain.

As Table 1.4 (on page 20) shows, repair and maintenance are of major importance and comprise nearly 50 per cent of the total annual activity – this includes all public and private sector work carried out on houses, infrastructure and commercial buildings. It is also evident that government departments and their agencies are significant clients of the construction industry. As both Tables 1.3 and 1.4 suggest, official statistics often draw a distinction between public and private sector activity. The public sector includes everything that is owned and/or funded by national or local governments such as roads, schools, the National Health Service, and local council leisure centres. In fact, a close examination of output tables reveals that approximately 20 per cent of construction industry turnover relates to public sector clients. Obviously this includes a vast range of contracts, varying in size from £10,000 for a small flood defence scheme to £5.3 billion for the new venues for the Olympic Games in east London.

Table 1.3 The construction industry – narrowly defined

Areas of Construction	Examples of type of work
Infrastructure	Water and sewerage Energy Gas and electricity Roads Airports, harbours, railways
Housing	Public sector/housing associations Private sector (new estates)
Public non-residential	Schools, colleges, universities Health facilities Sports and leisure facilities Services (police, fire, prisons)
Private industrial	Factories Warehouses Oil refineries
Private commercial	PFI (and similar public private partnerships) Schools/hospitals (where privately funded) Restaurants, hotels, bars Shops Garages Offices
Repair and maintenance	Extensions and conversions Renovations and refurbishment Planned maintenance

The percentage of public sector work in the UK has fallen considerably since 1980, as many of the activities traditionally in the public domain have been privatised. For example, privately owned utilities and services such as gas, electricity, water supply, telecommunications and railways were previously state owned activities. More recently, the private sector has been given a greater role in the funding, building and maintenance of public facilities such as hospitals, schools, prisons and roads. In these **public private partnerships,** the private sector organises the funds and manages the risks, while the public sector specifies the level of service required and ultimately owns the assets – as they are commonly returned to public ownership after 10, 15 or 25 years. These partnership arrangements are explained in Chapter 6. The important point for our purposes is that expenditure on the construction of public facilities – such as new schools and hospitals – is increasingly classified as private sector expenditure in the official data. (Figure 2.4 on page 38 shows the distribution of work across the public and private sector divide over the last 50 years.)

Table 1.4 Value of construction output in Great Britain

Types of work	Value of Output (£million) (Current prices)		
	1996	**2001**	**2006**
Infrastructure	6,337	7,147	6,532
Housing – public	1,422	1,438	3,442
private	5,592	8,797	19,572
Public non-residential	4,442	5,330	9,939
Private industrial	3,119	3,701	4,888
Private commercial	7,015	13,561	20,138
Repair and maintenance	27,317	34,720	49,058
Total (of all work)	**55,244**	**74,694**	**113,569**

Source: Construction Statistics Annual (BERR 2007: Table 2.1)

Sources of International Data

A *narrow definition* of the construction industry confines official statistical data to the 'site based' activities of firms involved with buildings and infrastructure. As Table 1.3 shows, this data is typically disaggregated into house building, private commercial building, infrastructure and civil engineering, repair and maintenance, and so on. Across Europe it is possible to see some common trends. Aggregating figures cross 28 European countries, 23 per cent of construction output is repair and maintenance, 27 per cent is house building, 20 per cent is infrastructure and 30 per cent is non-residential (FIEC 2007).

A modern alternative, however, is to widen the statistical definition and go beyond the narrow boundaries created by the international classification to include the whole life cycle of construction: design, production, use, facility management, demolition, etc. In fact, the Pearce Report (2003) argued that to fully understand the extent of what is meant by a sustainable industry required data relating to the broad scope of construction productivity including its environmental and social impacts. (The new approach adopted in this text will help to clarify the contributions that the sector makes to these wider concerns.) As Pearce (2003: 24) defined it, a *broad definition* should include the mining and quarrying of raw materials, the manufacture and sale of construction products, and the related professional services such as those of architects, engineers and facilities managers. For example, a detailed analysis of current available data in Great Britain indicates that in addition to the 1,500,000 workers employed in the traditional construction sector, there are a total of approximately 350,000 engineers, architects, facilities managers and chartered surveyors supplying professional services relating to construction and property, about 400,000 employed in the manufacture of building products and equipment,

24,000 involved in mining and quarrying construction materials and a further 600,000 workers selling construction materials and products. In effect this broad approach increases the significance of construction as the size of the industry virtually doubles in terms of employment and output. The implications of the contrast between the narrow and broad definitions will be reviewed further in Chapter 13 where we explain the measurement of economic industry activity in more detail.

The International Council for Research and Innovation in Building set up a project to analyse the entirety of the broad construction sector and published its findings in 2004. The new broad system approach was tested in nine countries and the project's conclusions confirmed the fact that the construction system as a whole seems to be roughly twice the size of the conventional construction sector (Carassus 2004: 190). For example, in Canada it was assessed that the broad construction cluster employed nearly 1,800,000 people, of which only 900,000 worked in the traditionally defined construction sector. Similarly in France the whole sector on a broad definition was assessed to be responsible for the employment of 2,358,000 people, of which 1,600,000 were employed on site as contractors (Carassus 2004: 45 and 78).

Table 1.5 Sources of international data

The European Construction Industry Federation (FIEC)	www.fiec.org

FIEC was created in 1905 to specifically promote the interests of the construction industry across Europe. The current membership is made up of 28 countries and 33 national federations. They represent 2.7 million construction enterprises employing a total of more than 15 million people. Publications include *Construction Activity in Europe* which gives a detailed statistical breakdown per member country.

United Nations Statistical Division	http://unstats.un.org/unsd/default.htm

This organisation is committed to the advancement of a global statistical system. Consequently it develops standards and norms for statistical activities, and supports countries' efforts to strengthen their national statistical systems. Comprehensive sets of data can be accessed and individual macroeconomic variables can be examined by selecting countries and years of interest.

Organisation for Economic Co-operation and Development (OECD)	www.oecd.org

The OECD encompasses 30 member countries sharing a commitment to the market economy (in some quarters the group is referred to as the rich countries club). Its work covers economic and social issues. Publications include *The OECD Observer*, a journal covering a number of themes including statistics and sustainable development, and a series of *Main Economic Indicators*, divided by subject (including construction) and by country.

Eurostat	www.ec.europa.eu/eurostat

Eurostat provides the European Union with a high quality statistical information service, and co-operates closely with other international data organisations (such as the UN and OECD). *The Eurostat Yearbook* is published annually and presents a comprehensive selection of statistical data covering areas such as labour markets, economy, international trade, industry, services and the environment (it is freely available online as a pdf file).

The problem faced by the academic researchers, however, was the different institutional arrangements that exist in each country and the different processes and systems of data collection. In other words, national sources of data usually conform to the established narrow definition of construction, as collated and presented by the European Construction Industry Federation (FIEC). But access and symmetry is lacking when attempts are made to measure the size of the industry using the broader construction system approach. Some of the inconsistencies will be uncovered if you follow the links detailed in Table 1.5, as in many instances a series of data cannot be precisely compared across national boundaries.

Sources of UK Data

Most of the key sources of economic data in the UK are published by the **Office for National Statistics** (commonly referred to as **ONS**). This is a government department, which reports to the Chancellor of the Exchequer. It was formed by the amalgamation of the Central Statistical Office (CSO) and the Office of Population Censuses and Surveys (OPCS) in April 1996. The Office for National Statistics publishes a *Guide to Official Statistics* – a comprehensive reference that signposts

Table 1.6 A brief guide to official sources of UK statistics

UK National Accounts

This publication is normally referred to simply by the colour of its cover as the Blue Book. It is published annually in the autumn by the ONS, and is considered to be a most important source of data for the UK macroeconomy, since it provides a comprehensive breakdown of GDP. As with other ONS publications, recent editions have become more user friendly. For example, there are now useful notes explaining how to interpret the accounts, a subject index and a glossary of terms.

Economic & Labour Market Reviews

A monthly publication covering issues broadly relating to the economy, the labour market and price inflation. Each issue contains a current economic review, independent forecasts and a mix of articles on various aspects of data. Its primary objective, however, is to provide a gateway to access the current statistical releases. The content always includes a few pages of key data plus a directory of online tables, to highlight web pages on the ONS site that have been recently updated. Much of the data in Part C of this text is drawn from this source.

Monthly Digest of Statistics

As the name implies, this is an amalgam of statistics; it has been published monthly since January 1946. It covers a wide range of topics, including economic, social and demographic issues. It presents data on prices, the balance of payments and employment.

Financial Statistics

Monthly publication of the ONS relating to financial indices such as interest rates, exchange rates and the money supply. Editions since 1997 also show data relating to inflation.

Bank of England Inflation Report

This is published quarterly with the Bank of England's *Quarterly Bulletin*. The *Inflation Report* serves a dual purpose. First, it provides a comprehensive review of specialised indices and a commentary on their forecasts. Second, it is the official publication in which the minutes of the monetary policy committee (described in Chapter 12) are made available to the public.

where to look for the different indicators. It is an excellent starting point to locate any data or sources not referred to in Table 1.6. It is revised periodically and most reference libraries have a copy of the edition published in 2000. The emergence of the ONS website and its search engine, however, has reduced the need for an update.

For specific information relating to construction, the key source in the UK is the *Construction Statistics Annual*. Compiled by the Department for Business Enterprise and Regulatory Reform (BERR), a new edition is published each year around August, and it can be freely viewed in its entirety on the BERR website (see page 30). The publication amalgamates all construction statistics produced by central and local government, trade associations and the private sector. It has an appendix that provides detailed notes on methodology and definitions to clarify the tables and figures. There is also a comprehensive subject index. Overall it provides a comprehensive picture of the UK construction industry through the last decade, together with some international comparisons. (For example, the references to industry sectors in Table 1.3 and data on the value of output shown in Table 1.4 are derived from *Construction Statistics Annual*.) As with most economic data there is always a time lag, so the 2007 edition only presents data up until 2006. Consequently, data quoted in this text can appear out of date before the book even goes to print. It is important, therefore, that you have the confidence to research data for yourself, and this is one of the reasons that website sources are provided in Table 1.5.

RESEARCHING DATA

When using official national statistics, in hard copy or from the Internet, it is useful to be aware of several conventions regarding their presentation. First, the symbols shown in Table 1.7 represent a summary of the main footnotes.

Table 1.7 Symbols used to annotate official statistics

●●	Not available
-	Nil or less than half the final digit shown
P	Provisional
R	Revised

These qualifying notes make it clear that published official statistics can be no more than an estimate. This observation is not made to discredit official statistics but to emphasise that any errors or omissions are corrected as soon as possible – the ultimate goal is to produce data sets that are as reliable and robust as possible.

Second, some statistical series do not have sufficiently consistent data to refer to the United Kingdom as a whole, and only refer to Great Britain or are restricted simply to England, Wales, Scotland or Northern Ireland. To give just one example, construction output figures derived from ONS data sets refer to the UK, whereas figures derived from BERR tend to be specific to Great Britain. There is also a possibility that the data set may only relate to a quarter – in which case, it may be necessary to multiply the figures by four to get an approximation for the whole year. So take great care when reading the headings and footnotes that are associated with each table.

Figures relating to money can be expressed in three ways.

- The most straightforward is to represent economic activity in 'face value' terms; or put more simply, in prices used for everyday transactions. Such measurements are referred to as **current prices**; an example is shown in Table 1.4.
- A more sophisticated option is to adjust figures to allow for inflation as this makes comparisons across time more meaningful, such adjustments are referred to as **constant prices**, as they are expressed in terms of a specific base year. As an example scan forward to Figure 2.4 (on page 38) where levels of construction activity over a period of 50 years can be confidently compared as the distortions caused by changes in value across time have been removed by referencing all prices to the year 2000.
- Time series data may also be expressed in terms of **index numbers** – in this type of data set the starting point (or base year) is given the value of 100, which allows subsequent percentage changes to be quickly identified. Index numbers and examples of their use are the subject of Chapter 14.

Finally, when reviewing any construction data it must always be borne in mind that the construction industry (right across Europe) comprises a very large number of small, geographically dispersed firms, and this makes it difficult for government agencies monitoring the industry to compile comprehensive data sets. Furthermore, alongside any official activities 'put through the books' and recorded in national statistics, there may be unofficial work carried out for 'cash in hand'. As a consequence, the data agencies in some countries include an estimated value for so-called 'unrecorded output'. The relationship between the size of the official (formal) economy and its unofficial (informal) equivalent is discussed further in Chapter 13.

Key Points 1.3

- ❑ The construction industry may be defined in many ways, but its impacts always cut right across the economy.

- ❑ A narrow definition of the construction industry includes the erection of buildings, infrastructure work, repair, maintenance and demolition. Examples of the activities are presented in Table 1.3.

- ❑ A broad definition of the construction industry adds to the narrow definition to include mining and quarrying of raw materials, the manufacture and sale of construction products, and related professional services. Examples of the broad range of activities are shown in Table 1.1.

- ❑ The annual value of construction output, in Great Britain, for the decade 1996 to 2006 ranged from £55 billion to £114 billion.

Reading 1

George Ofori's review of construction economics has become a standard reference point for those commencing a study of the subject. In fact, Andrew Cooke (1996: 13) goes as far as suggesting 'that all students of economics in a construction related field should read it at least twice – once before they embark on their studies and once when they are about to complete them'. The following extract provides an opportunity to partly meet his recommendation.

Unfortunately, as you will see, one of Ofori's main concerns is that there is no real consensus about the exact definition of construction economics. Yet since his paper was written, the aim of sustainable construction has gained momentum and the importance of improving economic efficiency in the industry has been broadly discussed in various government reports (e.g. Latham 1994; Egan 1998; HM Government 2008). This raises two related questions that you might consider now and as the text unfolds. First, what seems to have been the main hurdles in defining construction economics and, second, does the development of a sustainability agenda provide the necessary focus to create a more coherent approach?

The questions that relate to each of the readings are set to stimulate debate and further exploration. Those interested in pursuing the first two questions should see De Valence's (2006) paper where, across five pages, he manages to discuss the problems of classifying the boundaries of the subject area, and reviews most of the approaches that legitimately represent construction economics.

George Ofori (1994) 'Establishing Construction Economics as an Academic Discipline', *Construction Management and Economics* 12: 295–306

What is construction economics?

Perhaps the most basic feature of a discipline is a clear idea among its practitioners and researchers about what it entails, its aims and its boundaries. In this section, adopting a chronological approach, definitions of construction economics offered by various writers are considered.[1]

Drewer (1978) suggests that Turin was perhaps the first to attempt to impose scientific order on a 'pre-Newtonian' situation. Turin (1975) remarked that '...if economics is concerned with the allocation of scarce resources, it follows that building economics should be concerned with scarce *building* resources' (p. ix). Stone (1976) observes that 'building economics' embraces 'those aspects of design and production, and the related problems of organisation which affect the costs of a building' (p. xi), including forms of construction, methods of production, organisation of the industry and the impact of new methods, materials and forms of organisation and contractual relationships.

Rakhra and Wilson (1982) distinguish between 'building economics' and 'economics of building'. They suggest: 'Building economics takes an aggregate view of the building sub-section of the construction sector' (p. 51), embracing levels of building activity, the industry's contribution to the economy, impact of changes in government's policies and the nature, structure and organisation of the industry. 'Economics of building' was an 'examination at the specific project level of the resource transformation that is known as building' (p. 51), embracing cost-benefit consequences of design alternatives and choice of building components, life-cycle costing, effect of various combinations of labour and plant on site productivity and analysis of project resource requirements. Few writers refer to or adopt Rakhra and Wilson's (1982) distinction, notable exceptions being Bowen and Edwards (1985), Ofori (1990) and Bowen (1993).

Seeley (1983) opted for a very narrow definition, remarking that '...building economics has been widely used ... to describe the investigation of factors influencing building cost, with particular reference to the interaction of building design variables' (p. v). Ahuja and Walsh (1983) define cost engineering, which may be considered a form of construction economics, as '...an active approach in the design, construction and commissioning phases of a project, aimed at extracting the best possible value for money throughout each activity that has cost implications' (p. ix).

Hillebrandt (1985) adopted a broad perspective, defining construction economics as 'the application of the techniques and expertise of economics to the study of the construction firm, the construction process and the construction industry' (p. 1). Similarly, this journal defines construction economics as including design economics, cost planning, estimating and cost control, the economic functioning of firms within the construction sector and the relationship of the sector to national and international economics. Ashworth (1988) considers construction economics as embracing clients' requirements, impact of a development on its surrounding areas, relationship between space and shape, assessment of capital costs, cost control, life-cycle costing and economics of the industry in general.

Bon (1989), whose book was 'to offer a first step toward a theoretical framework for building economics' (p. xiii), suggests that 'building economics is about economising the use of scarce resources throughout the life cycle of a building...' (p. xiii) and concerns the 'application of standard investment decision criteria to buildings as a special class of capital assets' (p. xiii). Johnson (1990) adopts a similar definition, suggesting that '...knowledge of economics can provide a basis for making difficult trade-offs associated with both design and long-term management of buildings' (p. 9).

Ruegg and Marshall (1990) promise to show readers '...how to apply the concepts and methods of economics to decisions about the location, design, engineering, construction, management, operation, rehabilitation and disposition of buildings' (p. xi). Drake and Hartman's (1991) perspective is similarly project oriented, considering construction economics as being concerned with ends and scarce means in the Construction Industry' (p. 1057) and listing the, mainly surveying, techniques it embraces. Raftery (1991) suggests that 'building economics' could be said to be primarily about a combination of technical skills, informal optimisations, cost accounting, cost control, price forecasting and resource allocation. Finally, Bowen (1993) describes 'economics of building' as focusing 'on the application of quantitative techniques using financial criteria for the provision of financial advice to the design team' (p. 4).

From the above discussion, a common definition of construction economics does not exist. The chronological approach adopted helps to show that the issue has not become any clearer over time. For construction economics to develop into a discipline, a common definition is required to set the framework for issues to be considered and methodological approaches to be adopted. The definition should relate to the economic principles of scarcity and choice, refer to what is being studied (projects, practices, organisations and enterprises and industry) and state the overall aim of the discipline.

Segments

Two distinct segments of construction economics emerge from the discussion in the previous section. The first relates to construction projects, whereas the second concerns the industry. Ofori (1990) terms these 'construction project economics' and 'construction industry economics', respectively (these terms are used in the rest of the paper). However, again from the above definitions, some writers consider one or the other of the segments to be the entire field of construction economics or building economics. For example, Bon's (1989), Johnson's (1990) and Ruegg and Marshall's (1990) 'building economics' and Drake and Hartman's (1991) 'construction economics' relate only to projects and are similar to Rakhra and Wilson's (1982) 'economics of building' and Ofori's (1990) 'construction project economics', respectively. However, Stone's (1976) 'building economics' and Hillebrandt's (1985) 'construction economics' incorporate both segments. The present author prefers and adopts the title construction economics as a perspective encompassing both the project and the industry, as this enables all aspects of the field to be studied.

The project-related segment is basically about techniques (such as cost planning, life-cycle costing and value engineering) – Raftery (1991) likens it to 'cost accounting and management'. It is better known, as it has received greater attention from researchers and in course syllabi (Ofori, 1990). However, even here, there is some confusion. Seeley (1983, p. 1) appears to make 'cost control' synonymous with construction project economics and Kelly (1983), as well as Ferry and Brandon (1991), seem to equate 'cost planning' with construction project economics. Male and Kelly (1991) refer to cost management and define it as 'a synthesis of traditional quantity surveying skills ... with structured cost reduction/cost substitution procedures using the generation of ideas by brainstorming ... in a multidisciplinary team' (p. 25). Finally, Kelly and Male (1993) have combined cost management, which emphasises cost reduction at the design stage, with value management, which focuses on clients' needs prior to design, to obtain the 'comprehensive service' of project economics, which '...seeks to control time, cost, and quality during design and construction within the context of project functionality' (Marshall, 1993, p. 170).

It is necessary to delineate, agree upon and continuously research into and improve its segments and construction economics should be developed as an integrated whole.

Conceptual structure

Cole (1983) distinguishes between the core of a discipline, the 'fully evaluated and universally accepted ideas' (p. 111) found in all undergraduate textbooks and the research frontier which includes all on-going studies, most of which eventually turn out to be of little or no significance. Does construction economics have a core of confirmed and accepted concepts?

Some key terms

Precise and common definitions are indispensable building blocks in any discipline, a key base of its conceptual structure. Authors in construction economics often find it necessary to define their main terms (e.g. Batten, 1990; Ive, 1990). Bowen and Edwards (1985) and Bowen (1993) define such basic terms as estimating, forecasting, cost and price. Some construction economics terms, each of which has a clear definition in general economics, are considered in this section.

The industry

There is, as yet, no accepted definition of the construction industry (Ofori, 1990). Some writers consider it as involving only site activity, others include the planning and design functions and yet others extend it to cover the manufacturing and supply of materials and components, finance of projects or management of existing construction items (Turin, 1975; Hillebrandt, 1985). This leads to difficulties. For example, writers' basic data and inferences often differ, simply because they adopted different definitions of 'construction'.

Notes

[1] As you will see in Ofori's search for consensus he was prepared to include other authors references to building economics. He regards this as simply a narrower version of construction economics. As Ofori (1994: 296) explained at the start of the paper, 'of the two commonly used titles, building economics and construction economics, the latter embraces the former, covering also civil engineering and other forms of construction'.

References

Ahuja, H.N. and Walsh, M.A. (1983) *Successful Cost Engineering*, Wiley, New York

Ashworth, A. (1988) *Cost Studies of Building*, Longman, Harlow

Batten, D.F. (1990) Built capital, networks of infrastructure and economic development. In *Proceedings CIB 90 Joint Symposium on Building Economics and Construction Management,* Sydney, 14–21 March, 1, pp. 1–15

Bon, R. (1989) *Building as an Economic Process: An Introduction to Building Economics,* Prentice Hall, Englewood Cliffs, NJ

Bowen, P.A. (1993) A communication-based approach to price modelling and price forecasting in the design phase of the traditional building construction process in South Africa, PhD thesis, University of Port Elizabeth

Bowen, P.A. and Edwards, P.J. (1985) Cost modelling and price forecasting: practice and theory in perspective. *Construction Management and Economics,* 3, 199–215

Cole, S. (1983) The hierarchy of the sciences? *American Journal of Sociology,* 89, 111–39

Construction Industry Development Board (1988) *Construction Sector in a Developed Singapore*, CIDB, Singapore

Drake, B.E. and Hartman, M. (1991) CEEC: The organisation and activities of European construction economists. In *Management, Quality and Economics in Building,* Bezeiga, A. and Brandon, P.S. (eds), Spon, London, pp. 1056–69

Drewer, S. (1978) In search of a paradigm: notes on the work of D.A. Turin. In *Essays in Memory of Duccio Turin,* Koenigsberger, O.H. and Groak, S. (eds), Pergarmon, Oxford, pp. 47–51

Ferry, D. and Brandon, P.S. (1991) *Cost Planning of Buildings,* Blackwell Scientific Publications, Oxford.

Hillebrandt, P.M. (1985) *Economic Theory and the Construction industry,* 2nd edn, Macmillan, London

Ive, G. (1990) Structures and strategies: an approach towards international comparison of industrial structures and corporate strategies in the construction industries of advanced capitalist countries. *Habitat International,* 14, 45–58

Johnson, R. (1990) *The Economics of Building: A Practical Guide for the Design Professional,* Wiley, New York.

Kelly, J. (1983) Value analysis in early building design. In *Building Cost Techniques. New Directions,* Brandon, P.S. (ed.), Spon, London, pp. 116–25

Kelly, J. and Male, S. (1993) *Value Management in Design and Construction: The Economic Management of Projects,* Spon, London

Male, S. and Kelly, J. (1991) Value management and the economic management of projects. *The Building Economist,* 29(4), 25–30

Marshall, H.E. (1993) Value management in design and construction: The economic management of projects. *Construction Management and Economics,* 11, 169–71

Ofori, G. (1990) *The Construction Industry: Aspects of its Economics and Management,* Singapore University Press, Singapore

Raftery, J. (1991) *Principles of Building Economics.* Blackwell Scientific Publications, Oxford

Rakhra, A.S. and Wilson A.J. (1982) Building economics and the economics of building. *The Building Economist,* 21, 51–3

Ruegg, R.T. and Marshall, H.E. (1990) *Building Economics: Theory and Practice.* Van Nostrand Reinhold, New York

Seeley, I.H. (1983) *Building Economics: Appraisal and Control of Building Design Cost and Efficiency,* 3rd edn, Macmillan, London

Stone, P.A. (1976) *Building Economy: Design, Production and Organisation – A Synoptic View,* 2nd edn. Pergamon, Oxford

Turin, D.A. (1975) Introduction. In *Aspects of the Economics of Construction,* Turin, D.A. (ed.), Godwin, London, pp. viii-xvi

Extract information: Pages 298–9 of original plus relevant references from pages 304–6.

Part **A**

Effective Use of Resources

WEB REVIEWS: EFFECTIVE USE OF RESOURCES

On working through Part A, the following websites should prove useful.

www.berr.gov.uk

Following a change of Prime Minister in 2007, from Tony Blair to Gordon Brown, the DTI was disbanded and sponsorship of the construction industry was transferred to the Department for Business Enterprise and Regulatory Reform (BERR). According to its website, the department's objective 'is to secure an efficient market in the construction industry, with innovative and successful firms that meet the needs of clients and society while being competitive at home and abroad'. At the BERR home page it is possible to select a 'business sector' and navigate to construction, from here it is easy to download the *Construction Statistics Annual* and up-to-date information on sustainable construction. Both of these aspects were introduced in Chapter 1. Many of the other issues and reports referred to in the text are also supported.

www.competition-commission.org.uk
www.oft.gov.uk

Following the passing of the Enterprise Act 2002, there are two agencies responsible for monitoring market behaviour in the UK. The Competition Commission tends to focus on large firms concentrating on mergers and maintaining a level of competitive tension between companies; and the Office of Fair Trading (OFT) seeks to protect consumers against unfair market behaviour such as collusion. Their websites report on what is currently under enquiry, and provide summaries of reports, press releases, annual reviews and fuller information on their specific roles.

www.carol.co.uk
www.ft.com

Public limited companies must make their annual reports available to the public. These two sites facilitate the process. Carol is the acronym for Company Annual Reports On Line, and an interesting development in recent years is that many annual reports are now designed specifically for the Internet. Hard copies of company reports can also be requested by phoning the *Financial Times* – and www.ft.com provides the same opportunity online. The newspaper's website also offers unrestricted access to every story in the current edition and its vast archive. The site provides its own business search engine and the stated aim is to be the leading Internet resource for business people everywhere. It is certainly of use to students studying economics, especially as it is updated on a daily basis. It may be of particular use if you follow the advice in Chapter 8 and choose to study a specific construction firm.

www.ons.gov.uk

The Office for National Statistics is the UK government agency responsible for compiling, analysing and disseminating official statistics on Britain's economy, population and society at a national and local level. Detailed data are available from this site and it will enable the various statistical series used in this book to be updated.

2 Economic Systems for Resource Allocation

Decisions about resource allocation are necessary because we live in a world of scarcity. A review of the ideas listed at Key Points 1.1 and 1.2 should remind you of how central this basic premise is to the study of any branch of economics. To take a surreal example, when you open your front door in the early morning there are not millions of bottles of milk covering the neighbour's lawn; nor is there no milk. There is just enough bottled milk to meet the demand: say, the two pints your neighbour ordered. What this chapter seeks to explain is how this finely tuned allocation of resources can occur, given the multitude of construction, manufacturing and service resources that simultaneously need to be allocated.

The problems of **resource allocation** are solved by the **economic system** at work in a nation. In the case of construction, resource allocation has been strongly influenced by the public sector. *What* is produced, *how* it is produced and *for whom* can be determined by central government – and it was frequently – but governments across Europe now prefer to use a market system to answer the *what, how* and *for whom* questions. In general terms, therefore, it is possible to envisage two model systems. Each economic model brings together producers and consumers in different ways and each needs to be appreciated in order to understand how the universal questions about resource allocation are resolved.

Economic Systems: Two Extremes

The problem of resource allocation is universal as every nation has to tackle the issue of determining what, how and for whom goods and services will be produced. In Figure 2.1 we begin our presentation of the economic systems of the world by introducing two extremes: the **free market model** and the **centrally planned model** (along with two exemplar nations).

Figure 2.1 A spectrum of economic systems

On the extreme right-hand side of the diagram is the free market model, and on the extreme left-hand side, the centrally planned model. Cuba is a country whose system closely resembles the centrally planned model. At the other extreme is the USA, which comes close to the free market model. In between are the mixed economies of the remaining nations of the world.

FREE MARKET MODEL

The free market system is typified by limited government involvement in the economy, coupled with private ownership of the means of production. Individuals pursue their own self-interest without government constraints: the system is decentralised.

An important feature of this system is **free enterprise**. This exists when private individuals are allowed to obtain resources, to organise those resources and to sell the resulting product in any way they choose. Neither the government nor other producers can put up obstacles or restrictions to block those in business from seeking profit by purchasing inputs and selling outputs.

Additionally, all members of the economy are free to choose what to do. Workers may enter any line of work for which they are qualified and consumers may buy the goods and services that they feel are best for them. The ultimate voter in a free market, capitalist system is the consumer, who votes with pounds and decides which product 'candidates' will survive. Economists refer to this as **consumer sovereignty** as the final purchaser of products and services determines what is produced – and, therefore, 'rules' the market.

Another central feature of the free market economy is the **price mechanism**. Prices are used to signal the value of individual resources, acting as a kind of guidepost which resource owners (producers and consumers) refer to when they

Figure 2.2 The price mechanism at work

Price changes co-ordinate the decision-making processes of consumers and producers. When supply exceeds demand, the price of a product will need to fall for the market to clear. Conversely if demand exceeds supply, the price of the product will rise. For the price mechanism to function in all markets, it is important that resources are owned privately and can move freely between competing uses.

make choices. The flow chart in Figure 2.2 suggests how the price mechanism works. For example, when supply exceeds demand a price change occurs which brings the producers and consumers into harmony. This is precisely what happens during the January sales: the price of stock that has not been previously sold is reduced, to the point where demand is sufficient to clear the market. Conversely, when demand exceeds supply, the price of the good in question will rise until the market is in balance. This may be seen at a property auction where, to begin with, several buyers compete for a specific property: together they bid the price up, until finally there is only one interested party prepared to pay the final purchase price to the **vendor**.

Prices can thus be seen to generate signals in all markets (including factor markets): they provide information, they affect incentives and they enable buyers and sellers to express opinions. And, providing that prices are allowed to change freely, markets will always tend towards equilibrium, where there is neither excess demand nor excess supply. In short, a truly free market economy has the power to establish its own natural balance.

It may be self-evident by now that other terms used to describe the free market economy are 'market' or 'capitalist' economy.

Summary: What? How? For Whom?

What   In a free market economy, consumers ultimately determine what will be produced by their pattern of spending (their voting in the marketplace). As far as producers are concerned, their decisions about what goods to produce are determined by the search for profits.

How   Since resources can substitute for one another in the production process, the free market system must decide how to produce a commodity once society votes for it. Producers will be guided (by the discipline of the marketplace) to combine resources in the cheapest possible way to achieve a particular standard or quality. Those firms that combine resources in the most efficient manner will earn the highest profits and force losses on their competitors. Competitors will be driven out of business or forced to combine resources in the same way as the profit-makers.

For whom The 'for whom' question is concerned with the distribution of goods after production. How is the pie divided? In a free market economy, production and distribution are closely linked, because incomes are generated as goods are produced. People get paid according to their productivity; that is, a person's income reflects the value that the market system places on that person's resources. Since income largely determines one's share of the output 'pie', what people get out of the free market economy is based on what they put into it.

> ### Key Points 2.1
>
> ❏ We can illustrate and simplify the different types of economic systems by looking at two extremes: the free market model at one end of the scale and the centrally planned model at the other.
>
> ❏ The key attributes of a free market (model) economy are (a) limited government involvement, (b) individuals and producers express their desires through the price system, and (c) free enterprise – producers, consumers and resource owners all have complete freedom of choice over the range of products bought and sold.
>
> ❏ A free market economy is also called a market or capitalist economy.
>
> ❏ Any economic system must answer three questions. *What* will be produced? *How* will it be produced? *For whom* will it be produced?

THE CENTRALLY PLANNED MODEL

A centrally planned system (also referred to as a command economy) is typically characterised by a dominant government sector, coupled with the common ownership of resources. In other words, there is a central planning authority that takes the place of the price mechanism in allocating resources. The precise nature of a central planning authority depends upon the political system governing the economy. Indeed, it is worth noting that the terms 'socialist' and 'communist' properly refer to political systems and not economic systems. In fact, a right-wing dictatorship could operate a centrally planned economic system as effectively as a left-wing commune.

The common motivation for having a centrally planned system is the conviction that government commands are more likely to produce the 'right' mix of output, while the market mechanism may seem to operate in favour of the rich. Central planning creates the opportunity to direct resources to the society's most pressing needs. As the respected American economist J.K. Galbraith once observed, one of the few saving graces of the disintegrating communist economies is that nearly everyone has some kind of home, whereas many capitalist economies have not yet resolved the problem of providing affordable housing for the poor.

The flow diagram in Figure 2.3 outlines what a centrally planned system might involve. The three-stage process shown is a simplification of a bureaucratic reality. For instance, at stage one, various planning committees would exist to consider specific economic sectors and/or geographic areas. Similarly, at stage two, production targets and wages would be 'negotiated' with factory officials, workers, management and others involved in the chain of production. Finally, by stage three, the plans often become fraught with many difficulties, to the extent that there may be shortages and/or surpluses.

Figure 2.3 The general principles of a centrally planned economy

For such a system to work, the resources need to be centrally owned and controlled.

Stage 1

A state planning committee decides what the community needs as a whole

Stage 2

It sets production targets for firms at a fixed price and wage

Stage 3

The economy balances – all have jobs and basic goods

In today's market driven, global culture, such a system may seem hard to imagine, but it was used for more than 60 years from 1928 onwards when Stalin introduced Russia's first Five-Year Plan. For example, the State Planning Committee (Gosplan) in the former Soviet Union used to make sure that all construction projects identified the required quantity of labour, plant and materials. Then, all the plant and material requirements from projects due to come on stream in each region within a five-year period were added together. This enabled the planning committee to place the necessary orders with state contractors, material suppliers and plant production factories. Understandably this system was typified by cost estimating problems and over-employment, because in a planned economy there is no competition – no tendering – no risk and little chance of being laid-off work.

A similar system of resource allocation had been adopted by the People's Republic of China after its revolution in 1949. Following the death of the revolutionary leader Mao Zedong, the economic reforms of 1978 began the process of opening up trade links and allowing competition to operate. China is now fast becoming the workshop of the world and the Chinese clearly embrace the market system – although they still regard themselves as socialist.

As a consequence, most of the large, previously centrally planned, nation state economies are now involved in a transition towards market-oriented systems – and this will be reviewed in the next section on mixed economies. This transition is not without problems, however, as much is still influenced by the old regime of the centrally controlled plan. There is a kind of culture shock as a withdrawal of the 'visible hand' – as Rod Sweet (2002) refers to the planners and tweakers of a command economy – leads to the need to interpret the signals of the 'invisible hand' of the free market. In terms of the construction sector, this involves acquiring new skills to manage time, cost and quality.

Summary: What? How? For Whom?

What   In a centrally planned economy, the collective preference and wisdom of the central planners ultimately determines what is produced.

How   The central planners decide on the methods of production. This means that they need to know how many resources to allocate to each industry, many of which interrelate.

For whom   The relative rewards that people get are set by the central planners rather than the market. Thus market forces are not all important in determining factor rewards. There may be more opportunity to achieve some kinds of equality.

Key Points 2.2

❏ The centrally planned (model) economy relies on government commands rather than decentralised markets. It is also referred to as a command economy.

❏ The key attributes of a centrally planned (model) economy are (a) the government owns and controls most of the resources, (b) a planning committee identifies production targets, (c) the rewards for producing are usually set by the state rather than the market, and (d) there is a desire to achieve the 'right' mix of output.

❏ In any centrally planned system there are problems co-ordinating the different sectors.

❏ The difficulties of formulating and implementing the central plans have caused many command economies to introduce the market mechanism.

THE MIXED ECONOMY

The two economic systems introduced in this chapter do not exist in a pure form. The economic models used are simplified representations of the real world. In practice, most economic systems are far more complex; countries are neither purely free market nor purely planned. In the complex setting of everyday life, all nations have a **mixed economy**. Economists do not, therefore, study systems in which the activities of consumers and producers interact freely through a market or simply according to government plans. Economists study systems that contain mixtures of private decision-making and central organisation; the private decisions (being made in response to market forces) exist alongside the centralised controls of state legislation and economic plans.

One way of comparing the range of economic systems that presently exist in the world is to imagine them located on a wide spectrum (similar to that in Figure 2.1). In theory, each nation could be positioned according to the proportion of resources owned by the public and private sectors. The nations that have a high proportion of government owned resources would be located close to the centrally planned model, and those dominated by privately owned resources would be located at the other extreme near to the free market model. Such an exercise illustrates that all economies of the real world are mixed economies; some may come close to one or other of the economic models, but no nation fits precisely into the pure planned or pure free market category. During the last two decades there has been considerable movement along the spectrum. Indeed, we do not actually show the position of any countries in Figure 2.1, because as soon as the economies are located, their position seems out of date.

In general terms the **transition economies**, such as China and the states of the former Soviet Union, have shifted slowly away from the pure centrally planned model, while many European economies (particularly Britain) have moved closer towards the free market model under the influence of **privatisation** and **deregulation**. But transitions are rarely smooth, and there are various reports of economies struggling to maintain an equilibrium and steady rate of growth as they meet different cultural and economic hurdles. The United Nations *Economic Survey of Europe* carried out regularly since 1948 presents many mixed messages within its rigorous review of market and transition economies. In 2005 it reported that world output increased by some 5 per cent in 2004, and although this was the fastest average annual rate of increase in 30 years, the annual averages masked great extremes. Interestingly the closest example of an economy oriented toward the free market and the largest of those in transition, namely the United States and China, both provided the main stimuli to this global growth in 2004; whereas the rate of economic expansion in many of the other major economies had slowed down significantly (UN 2005: 19).

The transition of command economies since the Berlin Wall fell in November 1989 has attracted considerable coverage in both the news media and academic literature, as economies shifting towards free market systems reflect changes in national aspirations and culture. Both Russia and China, for instance, have tried to harness some of the efficiency of the market mechanism in an attempt to raise the living standards of the Russian and Chinese people and, consequently, the government's role in these nations has been reduced in recent years. These changes have not gone unchallenged. For example, China rejected restructuring if it meant destroying the socialist principles upon which the country was built.

In the UK, and elsewhere in Europe, the mix of public sector and private sector activity has varied over time because of the differing philosophies that the main political parties have adopted towards state intervention. For much of the post-war period the desirability of less or more public ownership of industry lay at the heart of the political divide in the UK between the Labour and Conservative parties. In terms of construction activity, from 1955 to 1985, much of the output was commissioned on behalf of a public authority such as a government department, housing association, or nationalised industry. For example, up until the 1980s the public sector was responsible for building council houses for those in need and providing a range of infrastructure services. In Figure 2.4 the outputs of construction (excluding repair and maintenance) across Britain, in constant (2000) prices, are plotted to show the divide between public and private sectors. Quite clearly up until the 1980s the public sector was responsible for a significant amount of the total activity. Construction work for railways, airports, telecommunications, ports, and electricity and water companies was commissioned by public sector authorities. In recent decades, however, the way that public capital is managed has undergone profound change, with both ownership and operation transferred to the private sector. This has been the case in many countries in the world, particularly in Europe. In Britain, the output of infrastructure has been separately recorded since 1980. Much infrastructure is owned and operated by the private sector – roads, footpaths, bridges, tunnels and parks are the odd ones out. There are, of course, some toll roads where users are charged and the providers aim to make a profit, but in the twenty-first century road infrastructure owned by the private sector is still something of an exception rather than the rule. In some cases, the line between the public and private sectors is blurred – but it would be safe to assume that water, sewerage, electricity, gas, telecommunications, railways, air transport and waterways are no longer the traditional domain of public sector management.

Figure 2.4 New construction output 1955-2006 (Great Britain)

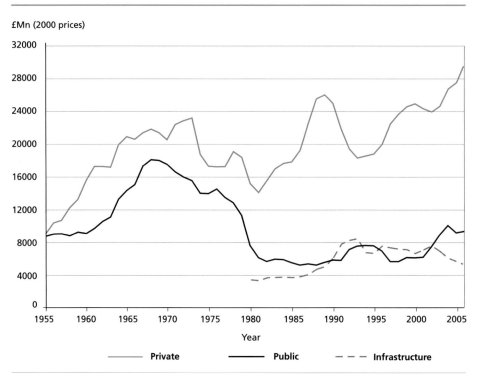

£Mn (2000 prices)

Source: Adapted from BERR (2008)

In other words economies have become strongly in favour of extending the principles of the market and the political divide has narrowed. Indeed, some kind of economic consensus based on market forces seems to be emerging and this has changed the political climate. As Samuelson and Nordhaus (2005: xvii) observe 'over the past quarter century, both attitudes and economic institutions have changed dramatically. Dozens of countries have rejected collectivist approaches and adopted market systems. Strong economic growth has been experienced in countries as diverse as Ireland, Botswana, and the Philippines.' Clearly, much can be attributed to the power of the market and the leitmotif for the foreseeable future seems to be an increasing reliance on unregulated private actions by individuals and companies in the hope that they will resolve the difficult questions of allocating resources efficiently.

Key Points 2.3

❏ In reality all economies are mixed economies, since elements of private markets and state coexist in all nations.

❏ It is the degree of market orientation, or of state intervention, that distinguishes one economic system from another.

❏ Economies of the world are in a state of flux and political views will continue to bring change.

❏ The current preference is for systems of resource allocation to be market led.

EQUITY, EFFICIENCY AND THE ENVIRONMENT

Much of the opposition that exists to market reforms relates to questions concerning the environment and the issue of social justice. There are certainly many environmental problems and social inequalities spawned by the market mechanism. For example, those with the greatest wealth have most 'votes' about what is produced, while those with no incomes, if left to fend for themselves, get nothing. On the other hand, the market mechanism does provide an efficient system of communication between producers and consumers that effectively signals 'what' to produce and 'how'. It does, however, fail to capture common goods such as environmental pollution.

Since the United Nations Conference on Environment and Development at the 1992 Earth Summit in Rio de Janeiro set the stage for sustainable development and related issues, the ideas and concepts lying beneath **equity, efficiency** and the **environment** have taken on an increasing significance. The basis of any sustainable development agenda, in any sector of the economy, must involve a mutual consideration of the business case, the wider community and the natural environment. Any sustainable outcome, satisfying the stakeholders representing these three interests simultaneously (whose needs often conflict) could genuinely be described as an 'elegant solution'. An important proposition of this text is that such a solution depends upon economics – as ultimately any policy that is fully sustainable boils down to questions of resource allocation.

Efficiency

In economics, efficiency is mainly concerned with resolving the questions of 'what' to produce and 'how'. The concept is accordingly often divided into two parts – **productive efficiency** and **allocative efficiency**, both of which are satisfied in a pure free market economy.

PRODUCTIVE EFFICIENCY

Productive efficiency means using production techniques that do not waste inputs. Expressed in the language of policy documents concerning sustainability, it means increasing growth rates while reducing the use of resources. In any free market economy businesses will never waste inputs. A business will not use 10 units of capital, 10 units of labour and 10 units of land when it could produce the same amount of output with only 8 units of capital, 7 units of labour and 9 units of land. Productive efficiency therefore refers to output that is produced at the lowest possible cost. During the last decade, for example, the increasing use of prefabricated components on site has enabled improvements in construction productivity – as this has encouraged greater levels of output with fewer workers. These developments depend upon managers responding 'correctly' to the various input prices facing them. The more expensive the inputs, the more incentive managers have to economise. The market signals, therefore, 'how' production should technically occur.

ALLOCATIVE EFFICIENCY

Allocative efficiency relates to maximising the total value (sometimes called utility) of the available resources. This means that resources are moved to their highest-valued uses, as evidenced by consumers' willingness to pay for the final products. The process of demand and supply guides resources to their most efficient uses. Individuals, as business people looking after their own self-interest, end up – consciously or unconsciously – generating maximum economic value for society. 'What' is produced, therefore, should involve no welfare losses; the utility of all groups in society should have been considered.

Equity

Equity does not, in its economic sense, simply mean equality. In this discipline, equity relates to fairness and social justice. From an economist's point of view, therefore, discussions of equity become closely related to considerations of sustainability such as respecting and treating stakeholders fairly – the 'for whom' question. Equity may also be broken down into two parts, **horizontal equity** and **vertical equity**, both of which depend upon government intervention.

HORIZONTAL EQUITY

This concept involves treating people identically. For example, a government policy of horizontal equity would support and promote equal opportunities between people of identical qualifications and experience, regardless of race or gender.

VERTICAL EQUITY

This concept is more contentious since it is concerned with being 'fair'. Vertical equity is about reducing the gap between the 'haves' and the 'have-nots'. It can involve governments providing targeted support to specific categories of people. For example, it may involve taxing the rich more heavily to provide services to support the poor.

It should be clear from these explanations that equity and efficiency often conflict. In fact, they can be imagined as polar extremes; as far apart as the two economic models shown in Figure 2.1. A centrally planned model economy, based on government intervention, is more able to support the concept of equity; whereas a free market model economy is able to foster a greater level of efficiency. It is the trade-off between these two qualities that, to some extent, accounts for the changes that nations are experiencing with their economic systems. The equity-efficiency trade-off certainly explains why all nations have become mixed economies. As *The Economist* (2002: 9) once commented (in an editorial on a UK budget): 'The vision is for a Britain which sits mid-way between America (free enterprise) and Europe (fairness), allowing business to grow fast while providing its people with good social services.'

Environment

Compounding the two way trade-off between efficiency and equity is the environment. The agenda of sustainability has significantly raised its profile. As the media continually reminds us, seas are being polluted, forests devastated, species are becoming extinct, rain is acid and the ozone layer is vaporising. These environmental issues are usually considered as **market failures** and will be dealt with fully in Chapter 10. They need to be introduced here, however, as they are central to understanding sustainability and implicated in *all* the resource allocation questions of 'what', 'how' and 'for whom'.

In economics, environmental goods and services are often discussed as a means of highlighting common problems of resource allocation regardless of the system. The explanation relies upon distinguishing between two categories of cost: **private costs** and **external costs**.

PRIVATE COSTS

These are the market costs of an individual's actions that are known and paid for directly. For example, when a construction business has to pay wages to site workers, it knows exactly what its labour costs are. When it has to buy materials or build a factory, it knows quite well what these will cost. Similarly, if individuals have to pay for car repairs, shoes or concert tickets, they know exactly what the cost (market price) will be. In short, private costs are those borne solely by the companies or individuals who incur them. They are 'internal' in the sense that a firm or household must explicitly take account of them.

EXTERNAL COSTS

These are not so straightforward, as they represent the costs of actions borne by people other than those who commit them. They are also referred to as third-party or neighbourhood costs. For example, consider the situation in which a construction business dumps waste products from its building site into a nearby river or in which an individual litters a public park or beach. Obviously, a cost is involved in these actions. When the construction firm pollutes the water, people downstream suffer the consequences. They may not want to swim in or drink the polluted water. The pollution may also ruin the prospects for fishing. In the case of littering, the people who come along after litter has cluttered the park or the beach are the ones who bear the costs. In other words, the creator of cost is not necessarily the sole bearer of that cost. The individual or firm does not internalise all costs – some external ones are overlooked.

Environmental problems may be described as situations in which the external costs exceed private costs. This is difficult for either market or command systems to control. The problem is that some collective resources are taken for granted – and the full costs of the economic activity are not accounted for, only the smaller private costs relating to some of the resources. This is neither efficient nor fair, and environmental problems can therefore be regarded as sitting midway on the spectrum of economic systems shown in Figure 2.5.

Figure 2.5 The trade-off between equity, efficiency and the environment

Adding the environment to the spectrum showing the trade-off between equity and efficiency brings in a third dimension which does not particularly lend itself to the market or command system. In effect, the problem is triangulated, and meeting all three criteria simultaneously is a difficult challenge.

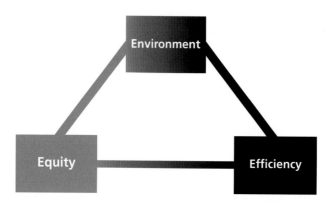

A society has to rely on public bodies such as the **National Audit Office**, or its equivalent, to assess the effectiveness of government spending and measure its impact on all resources. These activities assure that some kind of balance is achieved between economic efficiency, protection of the natural environment and the spending of taxpayers' money. Systems of **value management** can also be used to assess the efficiency, environmental impact and general effectiveness of projects. The use of value management is slowly increasing in both public and private sectors and will be discussed further in relation to construction projects in Chapter 6.

A recurring theme of this text involves studying how the construction industry can simultaneously embrace environmental, social and business concerns. Hence we shall, implicitly and explicitly, revisit the three strands of efficiency, equity and environment several times before reaching the concluding chapter of the book – which reviews the possibility of the construction industry becoming sustainable.

Key Points 2.4

- Ideas relating to efficiency, equity and the environment have gained validity since the sustainable development agenda has increased in importance.

- There are two concepts of efficiency: productive efficiency – when inputs are not wasted; and allocative efficiency – when resources are employed in their highest-valued uses.

- There are two concepts of equity (or fairness): horizontal equity – equality of opportunity; and vertical equity – actions to achieve social justice or fairness.

- A free market system encourages efficiency and a centrally planned system promotes equity.

- There are two concepts of cost: private costs – the costs to the first party; and external costs – the costs to third parties. The prevalence of environmental problems can be explained by the fact that external costs can be ignored or overlooked by those considering just private costs.

- Problems relating to the environment seem to pervade both free market and centrally planned systems of resource allocation.

- A recurring theme of this text is sustainable construction, which depends upon the industry embracing the three concerns of efficiency, equity and the environment.

3 The Market Mechanism

In Chapter 2 we emphasised that a central tenet of almost any modern economic system is a strong belief in the market mechanism (see Key Points 2.1 and 2.3). We also began to recognise that the forces of supply and demand are closely associated with this mechanism. This was highlighted in Figure 2.2. To interpret market signals perceptively – whether in the construction industry, building materials sector, property market or whatever – it is necessary to consider the price mechanism in more detail. We now begin to elaborate our understanding of the price mechanism. We can define it more fully as:

> an economic system in which relative prices are constantly changing to reflect changes in supply and demand for different goods and services.

The **price mechanism** is synonymous with the term **market mechanism**.

The forces of supply and demand are all encompassing within the market economy. Even before any good or service can be produced, the various factors of production need to be employed and their prices are affected by supply and demand. The forces of supply and demand bring together producers and consumers in such a way that appropriate goods and services are produced and appropriate incomes are rewarded. To highlight this interrelationship between supply and demand in different markets, we first consider an overview of a market-based economy.

ALLOCATION IN A MARKET ECONOMY

Figure 3.1 (see page 46) displays how the wishes of producers and consumers are 'signalled' by the price mechanism. To begin with **factor markets**: households (consumers) supply economic resources (factors of production) to firms that demand them to undertake productive activity. This is shown in the top half of Figure 3.1. The supply and demand of these factors (resources) determine the prices paid for them in the particular sector in which they are put to economic use. Householders receive differing levels of wages for their labour input, interest for the capital services they provide, rents for the land that they own and profits for their entrepreneurial abilities, according to the sector supplied.

In **product markets**, a similar scenario exists: firms supply various goods and services according to consumer demand. Again, it is the market price that balances the two parties' interests. This is shown in the bottom half of Figure 3.1.

Price Signals and Self Interest

For a market economy to function effectively, it is important that every individual is free to pursue 'self interest'. Consumers express their choice of goods or service through the price they are prepared to pay for them – in their attempts to maximise

Figure 3.1 Product and factor allocation via the price mechanism

In this simplified model households supply their services – labour, capital, land and entrepreneurial skills – to firms that demand them for production. The prices paid in factor markets are the balancing items that determine resource allocation. In product markets, firms supply goods and services to households that demand them, and prices provide the balancing item to determine distribution.

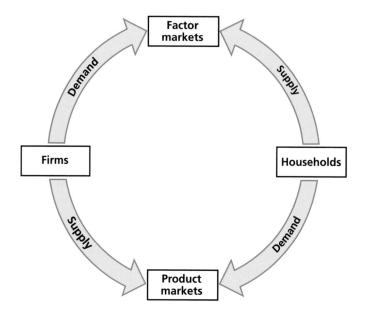

satisfaction. Producers, and owners of resources (and for most people this is their own labour power), seek to obtain as large a reward as possible in an attempt to maximise profit.

If consumers want more of a good than is being supplied at the current price, this is indicated by their willingness to pay more to acquire the good – the price is 'bid up'. This in turn increases the profits of those firms producing and supplying the good – and the incomes paid to the factors producing that good increase. As a result, resources are attracted into the industry, and supply expands.

On the other hand, if consumers do not want a particular product, its price will fall, producers will lose money and resources will leave the industry. This is precisely what happened in the 'new build' market during the early 1990s. The demand for new houses declined and prices fell; as a result, producers either concentrated on other construction work or went bankrupt. Consequently, between 1990 and 1993, the completion of new homes fell by over 50 per cent.

In simple terms, therefore, the price system indicates the wishes of consumers and allocates the productive resources accordingly. Or, in the terms we used in Chapter 2, the price mechanism determines *what* is produced, *how* it is produced and *for whom* is it produced. Now we shall examine how well these general principles apply to the UK construction industry.

Price Signals in the Construction Industry

According to the Department for Business Enterprise and Regulatory Reform there are more than 186,000 firms that make up the construction industry. This makes the market rather fragmented and very distinct from manufacturing. In the manufacturing sector there are usually a few firms producing a similar product that can be freely examined before purchase. In construction the opposite seems to apply – and a construction project usually involves many small firms combining their skills on site to produce a 'unique' specified product. This process means that the construction industry tends to be less efficient than manufacturing.

The reason individuals and businesses turn to markets to conduct economic activities is that markets generally reduce the costs of trading. These costs are called **transaction costs** because they are part of the process of making a sale or purchase. Economists define transaction costs as all of the costs enabling exchanges to take place. They include the cost of being informed about the qualities of a particular product, such as its availability, its durability record, its servicing facilities, its degree of safety, and so on. Consider, for example, the transaction costs in shopping for a new computer. These costs include phone calls or trips to sellers in order to learn about product specifications and prices, and subsequently there is the cost of negotiating the sale and, potentially, the cost of enforcing the contract should the machine fail to operate. In a purely theoretical or highly organised market these costs do not exist, as it is assumed that everybody has access to the knowledge they need for exchange to take place.

In the context of construction, however, these transaction costs are significant and go some way to explaining the adversarial nature of competition within the industry. A particular problem is the fact that often the product that is being created by a construction firm does not exist before the exchange is made. The product has to be precisely specified to assure the quality and quantity before an exchange can be agreed. And, complicating matters, firms will offer the product at differing prices. Consequently, contractual costs arise to clarify what is expected and to cover any contingencies in the proceedings, such as what happens if the work is handed over late, over price or fails to meet the anticipated quality. Gruneberg and Ive (2000) have identified six types of transaction costs that anyone demanding the services of a construction firm may have to meet. These transaction costs are summarised in Table 3.1 (see page 48).

Some economists argue that the increased use of electronic communication will reduce transaction costs both between businesses and between businesses and consumers. In fact, there is an ongoing academic debate about the significance of the **new economy** and its implications for B2B (business to business) and B2C (business to consumers) transactions. At present the construction industry does not effectively feature in this debate, as e-commerce has made little impact on the way construction firms communicate with one another or with their customers. High transaction costs, therefore, are destined to remain a hurdle to the efficiency of the modern construction industry working within a market context and we will revisit elements of this debate in Chapter 6.

Table 3.1 Transaction costs which affect construction

Search costs	The cost of having to find out what is on offer
Specification costs	To clarify precisely what is required to assure the desired quantity and quality
Contract costs	The cost of finding or creating conditions of contract to clarify what is expected and cover contingencies
Selection costs	The cost of choosing the best tender
Monitoring costs	To measure and control price, timing and quality
Enforcement costs	The legal bills relating to breached contracts

Source: Adapted from Gruneberg and Ive (2000: 123–4)

A belief in the importance of market signals dates back to the classical economists. They emphasised how prices and wages continually adjust to keep the general levels of supply and demand in balance. In fact, their belief in market forces was so strong that for a significant historical period economists recommended that the management of an economy could simply follow a laissez faire approach. Milton Friedman, a modern exponent of the market mechanism, strongly argued that the need for government intervention is minimal – as governments are only needed to provide a forum for determining the rules of the game and to act as an umpire to assure the rules are enforced. Interestingly, when this classical approach was first challenged, during the depression of the 1930s, construction was one of the first sectors that governments chose to manipulate.

Key Points 3.1

❑ The price mechanism is synonymous with the market mechanism.

❑ The forces of supply and demand in factor and product markets are reconciled by price (see Figure 3.1).

❑ Price movements provide the signals that freely responding individuals interpret, determining what is produced, how it is produced and for whom it is produced.

❑ Transaction costs are costs associated with exchange.

❑ Transaction costs are significant within the construction industry, and this reflects the fragmented and adversarial nature of firms involved in the process.

❑ Recognition of the importance of the market mechanism dates back to the classical economists.

GRAPHICAL ANALYSIS

Analysis of the price (market) mechanism has played a significant role within the history of economics. As far back as 1776, Adam Smith wrote in *The Wealth of Nations* about the 'hidden hands' of supply and demand determining market prices. Since then a standard part of any economics course has involved the study of supply and demand graphs. This is probably because it is easier to communicate an idea visually – as the saying goes 'a picture is worth a thousand words'. A supply and demand graph enables the relationship between price and quantity to be explored from the consumers' (demand) perspective and the producers' (supply) perspective. The standard layout of each axis is shown in Figure 3.2.

Figure 3.2 The axes of a supply and demand graph

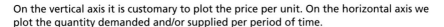

On the vertical axis it is customary to plot the price per unit. On the horizontal axis we plot the quantity demanded and/or supplied per period of time.

Quantity (supplied/demanded) per period of time

Using the labelled axis in Figure 3.2 can you determine what the pattern of demand in relation to the price would look like? To put the question in more formal terms, can you plot a **demand schedule**? Clearly, as the price of a commodity rises, the quantity demanded will decrease and as the price falls, the quantity demanded will increase. That is, from the demand side there is an **inverse relationship** between the price per unit and the quantity purchased: higher prices lowers (that is, cause smaller quantities of) demand. This is because consumers seek to maximise their satisfaction and get best value for money.

Using the labelled axis in Figure 3.2 can you determine what the pattern of supply in relation to price would look like? To rephrase the question in more formal terms, can you plot the **supply schedule**? Clearly, as the price of a commodity rises the quantity supplied will increase and as the price falls the quantity supplied will decrease. That is, from the supply-side there is a **direct relationship** between the price per unit and quantity sold: an increase in price usually leads to an increase in the quantity supplied. This is because suppliers seek to maximise their profit and get the biggest possible return for their efforts.

As suggested, these basic principles seem easier to appreciate when plotted on a graph. See if you agree by considering Figure 3.3.

Figure 3.3 A simple supply and demand diagram

The demand curve displays the fact that the quantity demanded falls as the price rises. The supply curve displays the converse relationship: that the quantity supplied rises as the price rises.

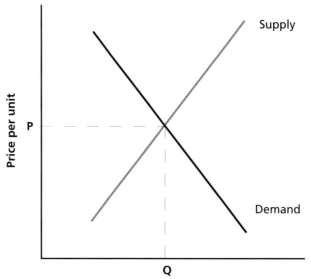

Quantity demanded & supplied per period of time

Three Qualifying Remarks

Economists have devised various methodological procedures to give their statements rigour and academic value. Three of these small but important techniques need to be highlighted especially as textbooks often fail to constantly enforce them.

PER PERIOD OF TIME

On the horizontal axis of Figure 3.2 and 3.3 we ended the statement with the qualification *per period of time*. This is to highlight that supply or demand is a flow that takes place during a certain time period. Ideally, the time period should be specified as a month, year, week, or whatever. Without a time dimension, the statements relating to quantity become meaningless.

CETERIS PARIBUS

The second qualifying remark relates to the Latin phrase **ceteris paribus**, which means other things being equal or constant. This is an important assumption to make when dealing with a graph showing two variables. For example, price is not the only factor that affects supply and demand. There are many other market conditions that also affect supply and demand and we cover these in Chapters 4 and 5. In the exercise above, constructing Figures 3.2 and 3.3, we assumed *ceteris paribus*. We did not complicate the analysis by allowing, for example, consumers' income to change when discussing changes to the price of a good. If we did, we would never know whether the change in the quantity demanded or supplied was due to a change in the price or due to a change in income. Therefore, we employed the *ceteris paribus* technique and assumed that all the other factors that might affect the market were held constant. This assumption enables economists to be more rigorous in their work, studying each significant variable in turn. The *ceteris paribus* assumption approximates to the scientific method of a controlled experiment.

SUPPLY AND DEMAND CURVES

When using supply and demand curves to illustrate our analysis, they will frequently be drawn as straight lines. Although this is irritating from a linguistic point of view, it is easier for the artist constructing the illustrations and acceptable to economists, since the 'curves' very rarely refer to the plotting of empirical data. It is worth noting, therefore, that so-called supply and demand 'curves' are usually illustrated as straight lines that highlight basic principles.

Key Points 3.2

- ❏ On a supply and demand graph, the horizontal axis represents quantity and the vertical axis represents the price per unit.

- ❏ Supply and demand curves illustrate how the quantity demanded or supplied changes in response to a change in price. If nothing else changes *(ceteris paribus)*, demand curves show an inverse relationship (slope downward) and supply curves show a direct relationship (slope upward) as shown in Figure 3.3.

- ❏ To understand the premise upon which supply and demand diagrams are drawn, it is important to remember three criteria: (a) the time period involved, (b) the *ceteris paribus* assumption and (c) the shape of the 'curves'.

THE PRICE IS RIGHT

Look again at Figure 3.3: inevitably there is a point at which the two curves must cross. This point represents the market price. The market price in Figure 3.3 is P and this reflects the point where the quantity supplied and demanded is equal, namely point Q.

At price P the market clears. There is no excess supply; there is no excess demand. Consumers and producers are both happy. Price P is called the **equilibrium price**: the price at which the quantity demanded and the quantity supplied are equal.

Most markets tend towards an equilibrium price, including the labour market, housing market, foreign exchange market, drainpipe market, or whatever. All markets have an inherent balancing mechanism. When there is excess demand, price rises; and when there is excess supply, prices fall. Eventually a price is found at which there is no tendency for change. Consumers are able to get all they want at that price, and suppliers are able to sell the amount that they want at that price. This special market concept is illustrated in Figure 3.4.

Figure 3.4 The determination of equilibrium price

In this example, the equilibrium price for each new flat is £100,000 and the equilibrium quantity is 2,000 units. At higher prices there would be a surplus: flats would be in excess supply and they will remain empty. For example at £150,000 the market would not clear; there needs to be a movement along the demand curve from H to E and a movement along the supply curve from h to E. These movements necessitate the price to fall. At prices below the equilibrium, there would be a shortage: flats would be in excess demand and there would be a waiting list. For example, at £50,000 the price would rise, reducing the demand from F to E and supply from f to E. This market will tend to settle at a price of £100,000.

The Concept of Equilibrium

The concept of equilibrium is important in economics and we will be referring to it in different markets and in different contexts as we study the economy. **Equilibrium** in any market may be defined as:

a situation in which the plans of buyers and the plans of sellers exactly mesh.

Equilibrium prevails when opposing forces are in balance. In any market, the intersection of a given supply curve and a given demand curve indicates the equilibrium price. If the price drifts away from this equilibrium point – for whatever reason – forces come into play to find a new equilibrium price. If these forces tend to re-establish prices at the original equilibrium point, we say the situation is one of **stable equilibrium**. An unstable equilibrium is one in which if there is a movement away from the equilibrium, there are forces that push price and/or quantity even further away from this equilibrium (or at least do not push price and quantity back towards the original equilibrium level).

The difference between a stable and an unstable equilibrium can be illustrated with two balls: one made of hard rubber, the other made of soft putty. If you squeeze the rubber ball out of shape, it bounces back to its original form. On the other hand, if you squeeze the ball made of putty, it remains out of shape. The former illustrates a stable equilibrium and the latter an unstable equilibrium.

Now consider a shock to the system. The shock can be shown either by a shift in the supply curve, or a shift in the demand curve, or a shift in both curves. Any shock to the system will produce a new set of supply and demand relationships and a new price-quantity equilibrium. Forces will come into play to move the system from the old price-quantity equilibrium to a new one. Now let us consider a specific example in the housing market.

A Change in the Conditions of the Market

To illustrate the dynamics of the market imagine what might happen if mortgage interest rates rise, while other things remain constant. This will reduce the demand for owner-occupied property at each and every price. This decrease in demand is shown in Figure 3.5 (see page 54) in the traditional economist way, by shifting the demand curve to the left from D_1 to D_2. If property prices now stay at P, consumers will only demand Q_a while suppliers (sellers) will continue providing Q. Consequently there will be an excess amount of supply in the market place equal to $Q - Q_a$. However, providing prices are allowed to move to make the amounts supplied and demanded equal again, suppliers will be able to off-load vacant properties by reducing their prices. As the price falls consumers will become interested in buying and demand will increase. Consequently a new equilibrium price will be arrived at. This new price is P_1 in Figure 3.5 and the new quantity being demanded and supplied will now be equal at Q_1.

The shifting of the demand curve (such as in Figure 3.5) only occurs when the *ceteris paribus* assumption is violated. In other words, the curves only shift to a new position when the market conditions change. We will explore these 'shifts' in more detail in the next two chapters.

Figure 3.5 Changing market conditions lead to a new equilibrium price

The leftward shift of the demand curve indicates that consumers are now willing and able to buy fewer properties in every price range due to the increase in mortgage rates. The excess supply of properties on the market at the old price P causes a new equilibrium to be found at the lower price P_1, at which the quantity demanded and quantity supplied are once again found to be equal.

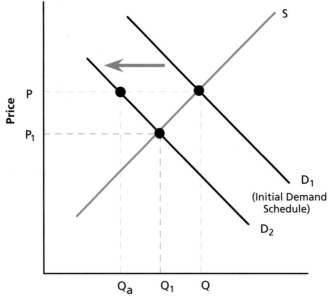

Quantity of property demanded and supplied for owner occupation per year

A CLOSING NOTE

Finally, to complete our introductory overview of market forces, we also need to recognise the existence of **market failure**. So far we have not assumed any market distortions. For example, we have assumed that labour will move freely to wherever work is most profitable and consumers will buy whatever they desire in freely determined markets. Yet, in reality, monopolies, oligopolies, subsidies, trade unions, externalities, high transaction costs and other market imperfections distort the situation. We shall be examining these issues in more detail in Chapter 10. But they are worth bearing in mind throughout the book, especially as our analysis is also concerned with how modern economies can manage to achieve construction that is sustainable, and these imperfections tend to create a gap between theoretical solutions and market-driven practice.

Key Points 3.3

❑ When we combine demand and supply curves, we find the equilibrium price at the intersection of the two curves. At the equilibrium price there is no tendency to change, the market clears (see Figure 3.4).

❑ Equilibrium exists whenever the separate plans of buyers mesh exactly with the separate plans of the sellers. Price points the buyers and sellers in the right direction.

❑ If conditions in the market change, the relevant curve shifts and a new equilibrium position is established (see Figure 3.5).

❑ Market failures exist, and market distortions form an important consideration when setting any agenda aimed at achieving sustainability.

4 The Theory of Demand

As suggested in the previous chapter and reviewed in Key Points 3.1 the concepts of supply and demand are the basic building blocks of economics, whether the outcomes are sustainable or not. In this chapter, we focus specifically on demand and in the next chapter we deal with supply.

When economists speak of demand they mean **effective demand**. Effective demand is money backed desire. It is *not* the demands of a crying baby or of a spoilt child wanting and grabbing at everything it sees. Demand from an economist's point of view is real, 'genuine' demand backed by the ability to make a purchase. It is distinct from need. For example, in 2008 the total number of households in England needing accommodation exceeded the total number of homes in the housing market. Only those who had sufficient means to 'demand' accommodation – that is, they could afford to buy or rent at market prices – were confident of securing somewhere to live. This anomaly explains the number of homeless people living in bed and breakfast hostels at a cost to the government.

Figure 4.1 A standard market demand curve

The demand curve for most goods and services slopes downward from left to right, as the higher the price, the lower the level of demand (other things being equal).

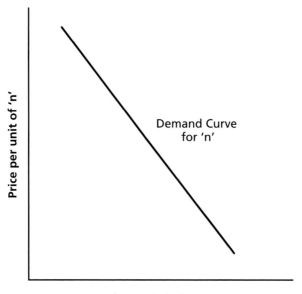

Demand Curve
for 'n'

Price per unit of 'n'

Quantity demanded of 'n'/month

THE BASIC LAW OF DEMAND

We already established in the last chapter (for example see Key Points 3.2) that a demand curve has a negative slope. It moves downward from left to right. This is illustrated in Figure 4.1.

The shape of the demand curve for most goods or service is not surprising when one considers the basic **law of demand**. This may be stated formally as:

at higher prices, a lower quantity will be demanded than at lower prices (and vice versa), other things being equal.

The law of demand, therefore, tells us that the quantity demanded of a product is inversely related to that product's price, other things being equal.

To continue with this analysis, we must consider the 'other things being equal' phrase more carefully. Clearly, demand is not only affected by price. As already implied in Chapter 3, it is easy to see that other factors may also affect demand. The demand curve shown in Figure 4.1 may shift to another position. Conditions in the relevant market may change significantly enough to cause consumers to change the quantity demanded at each and every price. For instance, imagine that Figure 4.1 represents the demand curve for a product of the construction industry. What events may cause more or less demand for that product at every price?

DEMAND IN THE CONSTRUCTION INDUSTRY

The determination of demand for goods and services produced by the construction industry is a complicated process. This is partly due to the size, cost, longevity and investment nature of the products and partly due to the broad range of what constitutes construction activity. This will become clearer as we consider factors affecting demand for various sectors of the industry.

Demand for Housing

At the beginning of the new millennium there were in excess of 26 million households in the UK. A large majority of these 26 million households could afford to demand a home, but it is important to remember the introductory point about effective demand and avoid confusion between 'need' and 'demand'. Each household requires some kind of shelter – a flat, bungalow, terraced house, maisonette, semi-detached house, cardboard carton, or whatever. The related resources are allocated through the market mechanism and the public sector, or through some mix of the two. It is difficult to envisage just one housing market. Indeed, when statisticians discuss the UK housing market it is usual to distinguish four sectors.

1 The *owner occupied sector* – that is, accommodation owned outright or being bought with a mortgage.
2 The *private rented sector* – property that is let at a market rent deemed 'fair' to tenants and landlords.
3 The *local authority rented sector* – housing made available by the local authority (council) at a subsidised rate from public funds.

4 The *registered social landlord sector* – which comprises non-profit making organisations that managing properties provided through public and private funds to rent to those in need. The sector is made up largely of properties managed by housing associations.

These four sectors have been listed in order of size according to the format used to officially measure tenure in the UK. In general terms, however, it is sufficient to understand that housing demand may be either for owner occupation or rent, or for some combination via shared equity. We now consider the main factors determining demand for housing within three main markets.

DEMAND FOR OWNER-OCCUPIED HOUSING

Most households in the UK, Ireland, Scandinavia, Australasia and much of Europe demand homes to own and occupy. Some 70 per cent of the households in the UK live in this form of tenure, and the relative size of the owner-occupied sector in Spain (80 per cent), Ireland (76 per cent) and Finland (78 per cent) is even larger. This form of ownership is generally supported by government initiatives that encourage demand by making the process of home buying as fast, transparent and as consumer friendly as possible. In the UK there have even been tax incentives for owner occupation, in the form of a subsidy on mortgage repayments offered regardless of income, though these have now been phased out. The logic is that if people own the property that they occupy, they will maintain it better. The feel-good factor derived from ownership makes the transaction costs of choosing and funding worthwhile, especially as a house provides an investment as well as a shelter. As you can imagine there are several factors that determine the demand for privately owned housing, and in Table 4.1 we identify the main ones.

Table 4.1 Factors affecting demand for owner-occupied housing

✓	The current price of housing
✓	The price of other forms of housing
✓	Income and expectations of change
✓	Cost of borrowing money and expectations of change
✓	Government incentives such as tax benefits
✓	Demographic factors such as the number of households
✓	Price of associated goods and services, such as maintenance, furniture, council tax, insurance, etc.

DEMAND FOR PRIVATELY RENTED HOUSING

Since the Second World War there has only been a small market demand in the UK for privately rented accommodation. At present approximately 11 per cent of UK households demand this type of accommodation. This is in direct contrast to some European economies where as many as 40 per cent of households are living in private rented accommodation. There has been a strong change in the pattern of housing demand in the UK. At the beginning of the twentieth century, people from all income levels routinely rented from private landlords. In 1915, for example, 90 per cent of UK families lived in the private rented sector. This change in the pattern of demand is closely associated with the supply drying up. This happened following the introduction of **rent controls** by government; these imposed a ceiling on rents – and created a big disincentive for landlords. As a result, current demand is mainly met by small-scale individual landlords who maintain and manage properties in their spare time. In an attempt to reverse this trend, the private rented sector in the UK has been largely deregulated in recent years.

In general, the market in private housing to rent varies greatly from country to country for a number of cultural and economic reasons. The main economic factors affecting the demand in this sector are listed in Table 4.2.

Table 4.2 Factors affecting demand for privately rented housing

✓	Current rent levels and expectations of change
✓	Income distribution – which determines affordability
✓	The cost of borrowing and expectations of change
✓	The law on rents and security of tenure
✓	Demographic factors, such as household formation
✓	The price of owner occupation

DEMAND FOR SOCIAL HOUSING

Housing provided by local authorities and registered social landlords, such as housing associations, is generally referred to as social housing. The origins of social housing lie in the idea that governments should pay a subsidy towards housing to make up for the shortage of accommodation available to low-income families. In the UK during the 1980s and 1990s – with both Conservative and Labour governments favouring free market policies – much of the local authority housing stock has either been transferred to housing associations to allocate, manage and maintain or sold to tenants thereby transferring stock to the owner-occupied sector. A similar process has been evident in the former Soviet Union, China, Czechoslovakia and Poland where privatisation of the social housing stock has been a key feature of the transition process.

Combined, the local authority and registered social landlords sectors still represent 'home' for approximately five million (18 per cent) of UK households. The factors that determine the demand for social housing are quite different from those driving demand for owner-occupied and privately rented housing. The main factors of demand for social housing are listed in Table 4.3.

Table 4.3 Factors affecting demand for social housing

✓	The current price (rent) of social housing
✓	The price level of other forms of tenure
✓	Assesment of need
✓	Availability of finance, such as income support and mortgages
✓	Levels of government subsidy

Demand for Industrial and Commercial Buildings

We now consider the demand for a whole range of buildings, including offices, factories, warehouses, hotels, garages, shops – in short, nearly all buildings except houses. Industrial and commercial buildings are not required for their own sake, but for the services they can provide. Consequently, demand for an industrial or commercial building is based on factors related to the specific sector in which the building will be used. Demand of this type is known as **derived demand**.

Derived demand implies that buildings are rented or purchased not because they give satisfaction, but because they can be used to produce goods or services that can be sold at a profit. This is different from the factors affecting the demand to buy a house. For example, in 2007 in the months following the emergence of problems in the US sub-prime mortgage market global business confidence was significantly dented and the demand for industrial and commercial property in the UK declined considerably.

Investments in industrial and commercial building, therefore, depend on the expectation that the buyers or renters – that is, businesses – will make profits in the future. If business confidence is low, investment will not take place – even if there is current demand for an increase in production or sales. The factors affecting demand for industrial and commercial buildings are largely dependent on the state of the economy, and business expectations concerning output and profit. In other words, because demand is derived, it is dependent on many things other than price. Some of the main factors of demand for this broad category of industrial and commercial buildings are shown in Table 4.4.

Table 4.4 Factors affecting demand for industrial and commercial buildings

✓	Technological developments
✓	Changes in taste or fashion
✓	Expected levels of cost, including interest rates
✓	The state of the economy and government policy
✓	Business confidence
✓	The age and condition of the existing premises

Demand for Infrastructure and Other Public Sector Construction

We now consider demands for major infrastructure projects such as hospitals, roads, schools, tunnels, prisons, museums, bridges, and police and fire stations. These demands are created by large numbers of individuals who, left to their own devices, are not able or willing to pay the market price for the desired facility. In these cases, the government decides the level of service that should be available. This does not mean that the public sector necessarily finances these major infrastructure projects. In many countries public policy, or a lack of public funds, means that the provision of these facilities has been transferred to the private sector. Examples of private sector provision include toll motorways on the Continent. In the UK, the private finance initiative (discussed in Chapter 6) ensures that many facilities previously provided by the public sector are now financed by the private sector.

The demand for these products is again largely derived. An excellent example comes from China where recent highway construction has taken place alongside the development of new cities, changing lifestyles and rapid economic growth. In other words, China does not want highways per se, but it demands the modern ways of transacting economic activity rather than rickety, congested local roads.

The demand for community oriented facilities can be judged on much the same basis as industrial and commercial buildings. However, assessing the demand for these products is even more complex as it also depends on the assessment of need and the funds available. Some of the main factors of demand for this broad area of construction activity are shown in Table 4.5. In considering these factors, one can understand the political difficulties of choosing which of the various public needs should be transferred into effective demand.

Table 4.5 Factors affecting demand for infrastructure and public sector construction

✓	Assessment of need – present and future
✓	Availability of finance and levels of government subsidy
✓	Government policy
✓	The age and condition of the existing stock

Demand for Repair and Maintenance

The demand for the repair and maintenance of buildings and infrastructure cannot be ignored, since it accounts for the biggest area of activity in the construction industry. In the UK, for instance, it represents around 50 per cent of all construction activity, and that does not account for the repair and maintenance activities carried out by the informal, black market economy which probably contributes significantly to meeting this type of demand.

Generally it seems that the poorer the country, in terms of GDP per capita, the lower the proportion of construction work is repair and maintenance. This can be explained by the fact that developed economies have greater stocks of construction products – so there is less need to add to it by new building, but there will be a greater need for repair and maintenance. Furthermore, as countries develop they often become concerned with maintaining and protecting their cultural heritage. In some cases, buildings are conserved even though it would be cheaper to knock them down and build anew. This trend may become more pronounced as the campaign for sustainability encourages the preservation of non-renewable materials.

As a result, there are very many factors determining the level of demand for repair and maintenance work, which embraces a wide range of activity. In general terms the key determinants are listed in Table 4.6.

Table 4.6 Factors affecting demand for repair and maintenance

✓	The current cost of repair and maintenance
✓	The cost of new building
✓	Level of current income
✓	Government policy associated with heritage and conservation, etc.
✓	The age and condition of existing stock (and its state of disrepair)
✓	The ownership pattern, since owner-occupied property tends to be better maintained

A GENERALISED DEMAND EQUATION

If we carefully study the factors affecting demand in the various sectors of the construction industry set out in Tables 4.1 to 4.6, we can quickly recognise some general themes. For example, it seems that recurring determinants of demand are the price of the good being considered, the price of related goods, the level of income and government policy. Therefore, we can quickly move towards stating a generalised tool for analysis.

Economic theory does not normally refer to just one sector of the economy. The analysis is undertaken in such a general way that it can equally apply to any sector. For example, the demand for any good is sometimes presented in the form of a general equation as follows:

$$Q_n^d = f\,(P_n,\, P_{n-1},\, Y,\, G, \ldots.)$$

This is formally referred to as a **demand function**. It may look complicated but it is only a form of shorthand notation. The demand function represents, in symbols, everything we have discussed above. It states that Q_n^d the quantity demanded of good 'n' is f a function of all the things listed inside the bracket: P_n the price of the good itself, P_{n-1} the price of other goods, Y income, G government policy, and ... a host of other things. (The notation relating to the basic demand function is set out in Table 4.7, see page 64.) These equations may be adapted and extended as necessary to support the analysis of a specific sector. For example, in our analysis earlier in this chapter, we were able to specify some of the other determinants – the other things – because we identified what 'n' specifically represented. For example, we often referred to the age and condition of the existing stock, and an assessment of need.

Table 4.7 Factors affecting the demand for any product

P_n	The price of the product
P_{n-1}	The price of other products
Y	Income
G	Government policy
...	A host of other things

Key Points 4.1

❑ The basic law of demand is that as price rises, lower quantities are demanded; and as price falls, higher quantities are demanded. There is an inverse relationship between the price and the quantity demanded, other things being equal (see Figure 4.1).

❑ In the construction industry there appear to be many determinants that affect demand, including the price (rent) of the building or infrastructure, the price of other goods (such as substitute goods), current level of income and government policy.

❑ Much of the demand for construction activity is of a derived nature, in as much as the goods are not necessarily demanded in their own right but for what they can add to the final good or service being produced.

❑ The relationship between the quantity demanded and the various determinants of demand can be expressed as an equation. The general terms used in the equation are shown in Table 4.7.

CHANGING MARKET CONDITIONS

Clearly there are many non-price determinants of demand, such as the cost of financing (interest rates), technological developments, demographic make-up, the season of the year, fashion, and so on. For illustrative purposes we will consider just four generalised categories – income, price of other goods, expectations and government – taking each in turn and assuming *ceteris paribus* in each case.

Income

For most goods, an increased income will lead to an increase in demand. The phrase increase in demand correctly implies that more is being demanded at each and every price. For most goods, therefore, an increase in income will lead to a rightward shift in the position of the demand curve.

Goods for which the demand increases when income increases are called **normal goods**. Most goods are 'normal' in this sense. There are a small number of goods for which demand decreases as incomes increase: these are called **inferior goods**. For example, the demand for private rented accommodation falls as more people become able to buy their own homes. (It is important to recognise that the terms normal and inferior in this context are part of an economist's formal language, and no value judgements should be inferred when the terms are used.)

Price of Other Goods

Demand curves are always plotted on the assumption that the prices of all other commodities are held constant. For example, when we draw the demand curve for lead guttering, we assume the price of plastic guttering is held constant; when we draw the demand curve for carpets, we assume the price of housing is held constant. However, the prices of the other goods that are assumed constant may affect the pattern of demand for the specific good under analysis. This is particularly the case if the other good is a **substitute good** (as in the example of guttering) or a **complementary good** (as in the carpet and housing example). Economists consider how a change in the price of an interdependent good, such as a substitute or complementary good, affects the demand for the related commodity.

Let us consider the guttering example a little more fully. Assume that both plastic and lead guttering originally cost £10 per metre. If the price of lead guttering remains at £10 per metre but the price of plastic guttering falls by 50 per cent to £5 per metre, builders will use more plastic and less lead guttering. The demand curve for lead guttering, at each and every price, will shift leftwards. If, on the other hand the price of plastic guttering rises, the demand curve for lead guttering will shift to the right, reflecting the fact that builders will buy more of this product at its present price. Therefore, a price change in the substitute good will cause an inverse change in the pattern of demand for the other alternative.

The same type of analysis also applies for complementary goods. However, here the situation is reversed: a fall in the price of one product may cause an increase in the demand for both products, and a rise in the price of one product may cause a fall in the demand for both.

Expectations

Consumers' views on the future trends of incomes, interest rates and product availability may affect demand – and prompt them to buy more or less of a particular good even if its current price does not change. This is particularly evident when we consider the demand for construction based activities. Consider the demand for housing (see Tables 4.1, 4.2 and 4.3). For example, potential house purchasers who believe that mortgage rates are likely to rise may buy less property at current prices. The demand curve for houses will shift to the left reflecting the fact that the quantity of properties demanded for purchase at each and every price has reduced due to consumer expectations that mortgage rates will rise.

Government

Legislation can affect the demand for a commodity in a variety of ways. For example, changes in building regulations such as the Code for Sustainable Homes, introduced in 2007, have placed a greater focus on the standards expected for water and energy usage. This, in turn, has influenced the design of, and demand for, the standard fittings and appliances used in kitchen, bathrooms and general heating. The demand for these products has increased, regardless of their present price. Or to put it another way, the demand curve for all code compliant products has shifted to the right, reflecting the fact that greater quantities of these units are being demanded at each and every price. (It will take some time for suppliers to adjust to these new standards, so prices might at first rise before technology catches up.) The government can also influence the level of demand by changing taxes or creating a subsidy. Tables 4.1 to 4.6 allude to the influence of government on construction demand, and this will be considered more fully in Chapters 9, 10 and 11 which review government policy relating to effective protection of the environment.

Revisiting Ceteris Paribus

When we first introduced the idea of holding other things constant, it may have appeared that these 'other things' were unimportant. The previous section, however, should have highlighted how wrong this interpretation would be. Indeed the *ceteris paribus* assumption enables economists to emphasise the fact that price and a host of other factors determine demand. Whenever you analyse the level of demand for any construction product there will always be a need to consider both the price and many other related factors. To clarify this important distinction between the price determinant and the non-price determinants, economists are careful to distinguish between them when they discuss changes in demand.

UNDERSTANDING CHANGES IN DEMAND

We have already explained that changes of non-price determinants cause the demand schedule to shift to the right or to the left, demonstrating the fact that more or less is being demanded at each and every price. These changes are often referred to as *increases* or *decreases* of demand.

Let us consider one example in detail. How would we represent an increase in the quantity demanded of naturally ventilated commercial buildings (at all prices) due to a respected piece of research concluding that air-conditioned buildings caused **sick building syndrome**? The demand curve for naturally ventilated buildings would shift to the right, representing an increase in the demand at each and every price. This is shown in Figure 4.2.

We could use a similar analysis when discussing decreases in demand due to a change in non-price determinants. The only difference would be that the demand curve would shift to the left, demonstrating that the quantity demanded is less at each and every price.

Figure 4.2 Change in a non-price determinant causing a shift in demand

If a non-price determinant of demand changes, we can show its effect by moving the entire curve from D to D_1. We assumed in our example that the move was prompted by some research in favour of naturally ventilated buildings. Therefore, at each and every price, a larger quantity would be demanded than before. For example, at price P the quantity of naturally ventilated buildings demanded increases from Q to Q_1.

By contrast, the price determinant causes a movement along the demand curve. This is obvious when one remembers that the demand curve represents price and quantity relationships. Changes to the quantity of demand due to price alone are often referred to as an *extension* or *contraction* of demand. This involves a move along the demand curve. When more is demanded at a lower price, this may be regarded as an extension from one coordinate on the demand curve to another. When less is demanded due to a rise in price, demand contracts. Such movements along the demand curve are described further in Figure 4.3 (see page 68).

Before we can begin to apply our theoretical knowledge of demand, it is particularly important to remember the distinction between a movement along, and a shift in, a demand curve. These rules will not only help us to understand the graphical analysis, but they will also enable us to acknowledge the numerous factors that come into play when interpreting demand in the construction industry.

Figure 4.3 Change in price causing a movement along a given demand curve

We show the demand curve for a hypothetical good X. If the price is P_1, then the quantity demanded would be Q_1; we will be at coordinate A. If the price falls to P_2, and all other factors in this market remain constant, then there will be an extension of demand to Q_2 – from coordinate A to coordinate B

Quantity of x demanded/year

ELASTICITY OF DEMAND

To complete this introduction to the theory of demand we need to briefly add that professional economists are also interested in measuring the degree of response of demand to a change in market conditions. The measurement of response is termed **elasticity**. It may be defined as:

> a measurement of the degree of responsiveness of demand to a change in an external variable.

A numerical value for the elasticity of demand may be calculated in relation to price, income or a related good using the formula:

$$\text{elasticity of demand} = \frac{\text{percentage change in demand}}{\text{percentage change in the explanatory variable}}$$

In this text we focus on price as the key variable. For instance, we may wish to know the extent to which a change in the price of petrol will cause the quantity of demand for petrol to change, other things held constant. Let's say that petrol prices rise by 10 per cent and this leads to a reduction in demand by 1 per cent. The calculation would be as follows:

$$\text{price elasticity of demand} = \frac{-1\%}{+10\%} = -0.1$$

This is a very small response and, in technical terms, whenever this measure is less than one demand is referred to as inelastic. The theory of demand states that quantity demanded is inversely related to the relative price, consequently price elasticity is always a negative number – if the price rises, which is a positive percentage change, the quantity demanded falls, which is a negative percentage change – but it is a convention that economists ignore the negative sign when discussing the price elasticity of demand. Therefore, if the price elasticity is stated as 0.3 it can be inferred that if price goes up by 1 per cent demand will go down by 0.3 per cent – so a 10 per cent increase will cause a decrease in demand of about 3 per cent. (By contrast, calculations relating to price elasticity of supply will nearly always be positive, as there is usually a direct relationship between the two – as price increases or decreases so does the quantity supplied.)

Another qualifying comment is the importance of percentage changes, as they allow us to measure price changes in terms of pence, pounds or thousands of pounds, and quantity changes in some other unit of measurement. The important thing is to be able to express the changes as relative values. A technical problem, however, arises from the way that percentage variables can be calculated. The standard way to compute a percentage change is to divide the change by the initial, or starting, level; but the same overall change will generate different answers depending on whether the starting point is the higher price and smaller quantity or the lower price and greater quantity. The way out of this difficulty is to take the average of the two prices and the two quantities over the range in question and divide the change with the average. This is known as the **midpoint method** of calculation, as it uses the average of the initial and final values to compare the percentage changes in price and quantity.

Even if you understand the mathematical rigour necessary to calculate the midpoint elasticity it is still not easy to be 100 per cent accurate. This is because changes in price are not the only thing to affect changes in quantity demanded: other factors, such as changes in income, population and technology also affect demand. So estimating the price elasticity of demand is problematic, but the basic model is useful, and it will be examined in greater detail in Chapter 5, when we discuss supply, as it will be easier at this point to introduce elasticity applications relevant to construction economics.

Key Points 4.3

❑ Elasticity of demand is given by the percentage change in quantity demanded divided by the percentage change in the explanatory variable (such as: price, income, or a related good).

❑ Price elasticity of demand is given by the percentage change in quantity demanded divided by the percentage change in price.

❑ The standard way to compute a percentage change is to divide the change by the starting point. The midpoint method calculates a percentage change by dividing a change by the average of the initial and final levels.

5 The Theory of Supply

In considering the theory of supply, we adopt a similar approach to that of the previous chapter on the theory of demand and, to some extent, this gives us an intellectual start. We will not, however, fully appreciate the suppliers' side of the market until we have worked through the next three chapters. Chapters 6, 7 and 8 are particularly significant. Chapter 6 explores the day-to-day relationship of contractors and clients and Chapters 7 and 8 deal with the theory of the firm.

Before formally commencing with supply it is useful to distinguish between cost and price. The basic difference arises from the perspective you are considering. For example, from a supplier's point of view, when a producer sells a good to a consumer the cost and price should not be the same. Normally, the producer seeks to make a profit – it is important, therefore, that the cost of the good is less than the selling price. Consequently, it is quite usual in construction for the cost of a project to be estimated and a mark-up for profits (risks) and overheads added before arriving at a price for the job. The contractor's mark-up is the difference between price and cost. In the present environment in the UK, however, many clients have become more knowledgeable and powerful; the client (or consultants acting on behalf of the client) predetermines an acceptable price and the contractor has to try to meet this figure.

The process is complicated further by the fact that for most construction work a price needs to be stated before the activity commences – when all the costs are not yet known. This contrasts with manufacturing: here, the producer does not have to determine the price until the activity is complete and all the costs have been revealed. Furthermore, it is important to understand that the most usual form of price determination in the construction industry is through some form of **competitive tendering**. This, in turn, makes it difficult for potential contractors supplying their services to take advantage of the market, as the lowest price bid is often seen as the most acceptable. According to Davis, Langdon and Everest (2000), in some countries the practice is to adopt the average or second lowest bid as clients are aware that competition based solely on the lowest possible price does not always represent best value. Similarly on large-scale public sector contracts in the UK it is becoming common practice to add a small percentage to the price of the winning bid to allow for the unrealistic cost estimates that may have occurred as ambitious contractors try to submit the lowest possible price at the bidding stage – this adjustment is known as **optimism bias**. Competitive tendering seeks to achieve fair comparisons on a like-for-like basis. What the subsequent chapters will highlight is the need for the efficient contractor to submit realistic tenders in relation to their costs and capacity.

THE BASIC LAW OF SUPPLY

We have already encountered the basic idea of supply in Chapter 3 and it might be useful to review Key Points 3.2 and 3.3. The supply curve slopes upwards from left to right, demonstrating that as price rises the quantity supplied rises and, conversely, as price falls, the quantity supplied falls. This is the opposite of the relationship that we saw for demand. The basic law of supply can be stated formally as:

> the higher the price, the greater the quantity offered for sale, the lower the price, the smaller the quantity offered for sale, all other things being held constant.

The law of supply, therefore, tells us that the quantity supplied of a product is positively (directly) related to that product's price, other things being equal. Or in terms of the discussion above, the number of potential contractors interested in bidding to supply a project will increase as the profit margin the client is prepared to meet rises. This is displayed in Figure 5.1.

Figure 5.1 The supply curve for an individual firm

The standard supply curve for most goods and services slopes upward from left to right. The higher the price, the higher the quantity supplied (other things being equal).

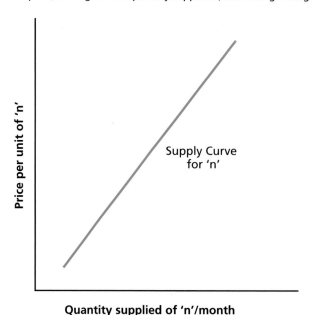

The Market Supply Schedule

The incentives within a specific market – and the constraints faced – are roughly the same for all suppliers. Each individual firm seeks to maximise its profits, and each firm is subject to the law of increasing costs. As we noted in Chapter 1 (see page 3), as a firm (or society) uses more and more of its resources to produce a specific item, the cost for each additional unit produced increases disproportionately. We referred to this as the law of increasing opportunity costs, and it highlights that resources are generally suited to some activities better than others. It is not possible to increase continually the quantity supplied of a specific item without cost increasing at a disproportionately high rate. In other words, when we utilise less well-suited resources to a particular production activity, more and more units of these resources have to be used to achieve an increase in output.

A firm's costs will also be affected by its fixed overheads. These vary according to the size of the firm and the nature of its activities. As we will explain in Chapter 7, the typical construction firm has relatively low fixed costs. The construction contractor's output is not based in a permanent factory, with all its related fixed costs. Each new construction site represents the firm's factory, and much of the fixed capital is hired as and when required.

We are now in a position to begin to appreciate the concept of a **market supply schedule**. The market supply of a product is given by the sum of the amounts that individual firms will supply at various prices. For example, at a price of £6 per unit, we might find that three firms are willing to supply 400, 300 and 200 units per day respectively. If these three firms make up the whole industry, we could conclude that at a price of £6 the market supply in this hypothetical industry would be 900 units per day. Let us consider this example in more detail. The relevant data is presented in Table 5.1.

Table 5.1 The individual and market supply schedules for a hypothetical three-firm industry

We see from the data that as price increases suppliers are willing to produce greater quantities. At the other extreme, low prices may actually discourage some firms from operating in the market. By combining the supply from each firm within the industry, we can identify the total market supply at each price; we do this in the final column.

Price	Quantities Supplied			
£/Unit	Firm A Units/day	Firm B Units/day	Firm C Units/day	Total Market Supply
4	0	0	0	0
5	300	0	0	300
6	400	300	200	900
7	500	380	250	1,130
8	580	460	280	1,320
9	620	500	290	1,410
10	650	520	295	1,465

In Table 5.1 we see how three firms comprising an industry perform individually at various prices. At low prices, producers B and C offer nothing at all for sale; most probably because high production costs constrain them. At higher prices, the law of increasing opportunity costs imposes constraints. By adding up each individual firm's output, at each specific price, we can discover the total supply that firms would be willing and able to bring to the market. We have highlighted the combinations at £6 per unit. As a brief educational exercise you could try to plot the market supply schedule on a graph. If you do it correctly, the supply curve for the market should be similar to that for an individual firm – a curve sloping upwards from left to right, as represented in Figure 5.1.

Key Points 5.1

❏ Before a final price to supply work can be determined a construction contractor needs to carefully estimate costs and assess an acceptable level of profits.

❏ The basic law of supply is that as price rises larger quantities are supplied, and as price falls smaller quantities are supplied. There is a direct, or positive, relationship between price and the quantity supplied, other things being constant (see Figure 5.1).

❏ The size of each individual firm will determine how much it can produce at various prices. Fixed overheads and the law of increasing opportunity costs affect each firm differently.

❏ The market supply of a product is derived by the summation of the amounts that individual firms will supply at various prices. Plotting these total amounts against their related prices enables one to construct a market supply curve.

SUPPLY IN THE CONSTRUCTION INDUSTRY

Many firms contribute to the supply of construction products, including large national contractors, material manufacturers, plant hirers and local site labourers. So while it may be theoretically possible to estimate construction supply by summing what the firms in the market are willing to supply at various prices, the huge range of private contractors involved in construction complicates the process of simply aggregating individual supply curves. There are approximately 186,000 firms in the UK supplying construction-based activity and, as we have discussed in earlier chapters, the industry is clearly not one simple market. There can be little competition in the supply of products between the local builder undertaking repair and maintenance in a small town and a large national civil engineering firm. They supply separate markets. The important point to note is that we need to consider factors affecting supply in specific sectors of the industry. This mirrors the approach taken in the previous chapter on demand. We need to envisage the construction industry as several different markets, each with distinct factors that affect supply and demand.

The total supply of construction output in Great Britain is broken down into specific activities according to monetary value in Table 5.2. House builders supply approximately 20 per cent of the annual output. Contractors supplying industrial and commercial buildings account for 22 per cent of the total output in value terms. The large building and civil engineering firms undertaking complex infrastructure projects such as motorways, power stations, airports and other public sector activity relating to education, health and the police account for a further 15 per cent of the industry's supply. This leaves the large number of small general builders dealing with repair and maintenance contracts to supply the lion's share of the activity with 43 per cent.

Table 5.2 Construction industry supply in Great Britain, 2006 (at current prices)

Type of work	Value (£million)	Percentage of total
New public housing	3,442	
New private housing	19,572	
Total	23,014	20.3
Private industrial	4,888	
Private commercial	20,138	
Total	25,026	22.0
Infrastructure	6,532	
Public (non residential)	9,939	
Total	16,471	14.5
Total repair and maintenance	49,058	43.2
Grand total	**113,569**	**100.0**

Source: Construction Statistics Annual (BERR 2007: Table 2.1)

It should be pointed out that the labour, capital and management resources employed on any one construction project could transfer to another. In fact, it is quite common for a specialist trade firm to move from site to site as contracts are fulfilled, and few firms remain in place for the whole duration of a project. It is this overlapping nature of the sectors comprising construction that give rise to some common reference points for factor rewards across the industry. In other words, rates of profit, wages and material prices tend towards some kind of equilibrium. Indeed, as competition intensifies or diminishes in particular sectors across the industry, suppliers could decide to shift the use of their resources to gain higher rewards. This will become clearer as we discuss in subsequent chapters patterns of cost and contracting characterised by differing levels of competition.

SUPPLY AND THE PRICE DETERMINANT

The law of supply states that more goods are supplied at higher prices, other things being held constant. This is because at higher prices there is greater scope for firms to earn a profit. Firms already in the market have an incentive to expand output, while higher prices may also enable those firms on the fringes of the market to enter the industry. At higher prices, therefore, the increased quantity supplied is made up by existing firms expanding output and a number of new firms entering the market. For example, in Table 5.1 we showed that in our hypothetical industry at a price of £5 per unit market supply was 300 units per day, but higher prices enticed other firms into the market and total supply increased.

SUPPLY AND NON-PRICE DETERMINANTS

Up until now, we have discussed supply and its related curve on the assumption that only price changes. We have not effectively considered any other determinants that influence producers' behaviour. We have constantly reiterated the *ceteris paribus* qualification, that other things are held constant. Some of these 'other things' assumed constant are the costs of production, technology, government policy, weather, the price of related goods, expectations, the goals of producers (do they wish, for example, to maximise profits or sales), and so on. Now, we shall broadly consider four of these non-price determinants.

Cost of Production

We have implied that producers are seeking to maximise their profits. Therefore, any change in production costs will, *ceteris paribus*, affect the quantity supplied. To illustrate this principle, return to Table 5.1. If unit production costs increase by £1, and this additional cost cannot be passed on by suppliers, then they will supply less to the market at each price. These changed conditions will cause the market to shrink so that, for example, only 300 units per day would be supplied at a price of £6 per unit.

In technical terms, what is happening is that the supply curve has shifted to the left: less is now supplied at each and every price. The opposite would occur if one or more of the inputs became cheaper. This might be the case if, say, technology improves, but such opportunities seem slow to emerge in a construction industry that is both labour intensive and culturally inclined to invest little in research, development and training.

Government

In a similar way, taxes and subsidies also affect costs and thus supply. For example, the landfill tax has increased construction costs and reduced supply at each price. A subsidy would do the opposite, and increase supply at each price, since every producer would be 'paid' a proportion of the cost of each unit produced by the government. A more complicated issue is the impact of general taxation, as much of the construction industry's demand is derived and depends on how others forecast their requirements. The most direct impact that the government has on construction markets is through legislation. Obviously the industry is affected by changes in statutory regulations that apply to building, planning, and health and safety.

Supply Chain Management

It would be a very rare for a contracting firm to be able to complete any construction activity entirely alone – just consider, for example, the variety of materials that need to be supplied for a typical project. Most construction activity normally involves integrating and managing a whole host of activities and processes to reach the final product, including subcontracting skilled work and purchasing materials. One medium-sized contractor with an annual turnover of £240 million (made up of contracts averaging £7 million each) claimed that 75 per cent of its total work was subcontracted, 8 per cent was spent on materials and 4 per cent was used to hire plant (Jessop 2002: 7).

The larger firms, especially the huge conglomerates, ensure their clients are provided with prompt and reliable services by diversifying into other businesses to extend their range of operations; for example, a construction firm may choose to merge with its material supplier to guarantee it meets completion targets on time. Such a merger would also eliminate many of the associated transaction costs. In some recent procurements (which we discuss in Chapter 6), the contractor has even agreed to take a stake in the completed project.

This discussion emphasises that construction firms not only produce different products but they also operate outside their immediate business and, to understand the supply implications, we find ourselves considering changes in many related markets as well as the conditions in the construction industry.

Expectations

A change in the expectations about future prices or prospects of the economy can also affect a producer's current willingness to supply. For example, builders may withhold from the market part of their recently built or refurbished stock if they anticipate higher prices in the future. In this case, the current quantity supplied at each and every price would decrease, the related supply curve would shift to the left.

Key Points 5.2

❑ Supply within construction is made up of several interrelated markets. Output is determined by thousands of firms that can transfer to other sectors of the industry if they think it would be worthwhile.

❑ The supply curve is plotted on the assumption that other things are held constant. Four important non-price determinants are (a) costs of production (including technological changes), (b) government, (c) supply chain issues and (d) expectations.

UNDERSTANDING CHANGES IN SUPPLY

Just as we were able to distinguish between shifts of, and movements along, the demand curve, so we can have the same discussion for the supply curve. A change in the price of a good itself will cause a movement along the supply curve, and be referred to as an *extension* or *contraction* of supply. A change in any non-price determinant, however, will shift the curve itself and be referred to as an *increase* or *decrease* in supply.

Let us consider one example in detail. If a new computer-assisted design and cost estimating package reduces fees relating to new builds, then design and build contractors will be able to supply more new buildings at all prices because their costs have fallen. Competition between contractors to design and build will ultimately shift the supply curve to the right, as shown in Figure 5.2. By following along the horizontal axis, we can see that this rightward movement represents an increase in the quantity supplied at each and every price. For example, at price P, the quantity supplied increases from Q to Q_1. Note that if, on the other hand, the costs of production rise, the quantity supplied would decrease at each and every price and the related supply curve would shift to left.

For analytical purposes, it is helpful to distinguish the cause of changes in supply. In our example about computer-assisted design, it would have been wrong to conclude that price has simply fallen and quantity supplied expanded accordingly. The reason for the increase in supply, at all prices, is due to a change in technology.

Figure 5.2 A shift of the supply curve

If price changes, we move along a given curve. However if the costs of production fall, the supply curve shifts to the right from S to S_1 representing an increase in the quantity supplied at each and every price.

Quantity of new builds supplied per period

Elasticity

We introduced the concept of elasticity in Chapter 4 (see Key Point 4.3 for a summary). Economists are often interested in the degree to which supply, or demand, responds to changes in market conditions. For example, the measurement of price responsiveness is termed **price elasticity**. It may be defined as:

a measurement of the degree of responsiveness of supply to a change in price.

A numerical value for the **price elasticity of supply** (PES) is calculated using the same midpoint approach outlined in the elasticity of demand section on page 69. In this instance, the formula is:

$$PES = \frac{\text{percentage change in quantity supplied}}{\text{percentage change in price}}$$

What the formula tells us is the *relative* amount by which the quantity supplied will change in relation to price changes. For example, if a 10 per cent increase in price leads to a 1 per cent increase in the quantity supplied the price elasticity of supply is 0.1. That is a very small response. There are in effect three types of measure that economists use as a reference point to discuss price elasticity.

1   **Price-inelastic supply**

When the numerical coefficient of the price elasticity of supply calculation is less than 1, supply is said to be 'inelastic'. This will always occur when the percentage figure for the change in supply is smaller than the percentage figure for the change in price. A PES coefficient of anything between 0 and 1 represents a situation of inelastic supply. The introductory example in which a 10 per cent increase in price led to a very small response in supply suggests a price-inelastic response: the measured coefficient was 0.1. In most cases where firms are supplying into the construction industry the price elasticity of supply, in the short term at least, will be inelastic.

2   **Price-elastic supply**

When the numerical value of the price elasticity of supply calculation is greater than 1, supply is said to be 'elastic'. This will always be the case when we the percentage change in supply is larger than the percentage change in price. For example, if a 5 per cent rise in price leads to a 50 per cent increase in quantity supplied, the PES coefficient will be 10. In other words, a small change in price elicits a large response in supply. This would be unusual occurrence in the markets for construction or property – but not impossible.

3   **Unit-elastic supply**

This is the most hypothetical case, as it describes a situation in which a percentage change in price leads to an identical percentage change in supply. This will always produce a coefficient value of 1, since the same figure appears on both the top and bottom lines of the price elasticity of supply formula.

Elasticity calculations played a significant part in the review of housing supply undertaken by Barker for the UK government in the early 2000s. For example, by drawing on several comprehensive academic estimates of the price elasticity of supply of housing in the UK carried out between 1974 and 2003, there appeared to be a considerable degree of consensus that UK housing supply is relatively unresponsive to price changes, with output increasing by proportionately less than price. The estimates taken from different time periods suggested that the price elasticity of supply of housing is always less than 1. Most estimates tended to be in a range between 0.3 and 0.8 (Barker 2003: 42).

Much of Barker's (2003) work examined the way households and house builders react to changing circumstances, such as rising incomes and price changes. For example, it was suggested that UK households have a high propensity to consume housing services as incomes rise over time. These are the comparative income elasticities estimates calculated by the Treasury for various European countries that were presented as evidence: Germany 0.0; Italy 0.4; Finland 0.5; France 0.6; Sweden 0.7; UK 1.0; Belgium 1.0; Ireland 1.1; Netherlands 1.2; and Spain 1.9. Compared to many other European countries, the UK has a high income elasticity of housing demand. But this position is complicated by the fact that the UK also has a low responsiveness of demand to house price changes. A low price elasticity of housing demand means that a one per cent rise in house prices results in less than a one per cent fall in spending on housing. The Barker study estimates that the UK's price elasticity of demand is 0.5. These factors, combined with low elasticity of supply in response to price changes, suggests that the UK's housing market is set on a path of long-term rising prices (Barker 2003: 39).

PRICE ELASTICITY OF SUPPLY AND TIME LAGS

For our purposes it is sufficient to understand that in the short run the price movement of construction goods tends not to affect supply. To increase the supply of any good or service takes time. If firms have some surplus capacity, they may be able to increase production fairly rapidly, but once they reach full capacity supply is fixed until extra capacity can be installed. For construction this is a particular issue – and it is common to talk about short-run and long-run supply. The **short run** is defined as the time period during which full adjustment – to, say, a change in price – *has not* yet taken place. The **long run** is the time period during which firms have been able to adjust fully to the change in price.

In the short run, rental values and house prices are demand determined because adjustments cannot quickly be made to the supply of property. The markets for construction in the short term are price inelastic in supply. In fact, it is the inelastic supply relative to demand that causes property markets to be unstable and characterised by fluctuating prices. In the extreme short run, the supply of buildings or infrastructure is fixed and the supply curve is a vertical straight line. Such a scenario is shown in Figure 5.3. It shows that the quantity supplied per year is 100,000 units regardless of price. For any percentage change in price, the quantity supplied remains constant. Look back at the formula for calculating elasticity. If a *change* in the quantity supplied is zero, then the numerator is zero, and anything

Figure 5.3 Perfectly inelastic supply

The supply curve is vertical at the quantity of 100,000 units per year, as the price elasticity of supply is zero. Producers supply 100,000 units no matter what the price.

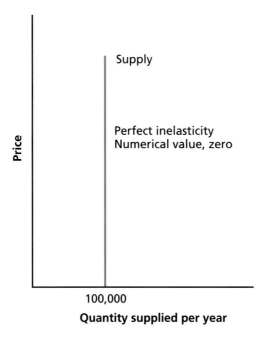

Quantity supplied per year

divided into zero results in an answer of zero. This is defined by economists as **perfect inelasticity.** (Note, exactly the same situation can be envisaged for demand: that is, a vertical demand curve represents zero elasticity at every price, too. For example, the price of electricity or gas may increase but in the short term consumers will continue to demand the same amount of energy until they have had time to switch over to more energy efficient options.)

Time tends to be the main determinant of the elasticity of supply. In the immediate time period supply is fixed, and inelastic to the value of zero; but given time for adjustments, supply increases can be organised and responses become elastic. This feature of supply inelasticity is particularly notable within property markets. Land is characterised by being perfectly inelastic; that is, as property prices increase the quantity supplied does not alter. Undeveloped areas can be developed and existing areas of land can change use, but both these possibilities take time.

So there will always be a time lag between price signals and changes to output in the construction industry. The duration of the delay is determined by a number of factors including site identification, site acquisition, negotiating planning permission, and the period of time needed for construction. These factors are examined in more depth in the Barker report (2003), where it is confirmed that the supply elasticity will be higher in the longer term as companies have time to change their factors of production and/or methods of construction. As Barker (2003: 41) specifically points out, in the very short run there is some scope to respond to changes in demand, as builders currently building a phased development could increase temporarily the build rate on that site (though this might be at the expense of less output in the next time period if the supply of land does not accelerate). Furthermore, in principle it should be possible to switch land from one use to another. However, as parcels of land in any region or area become scarcer, opportunities for substitution become extremely limited. This is particularly important given the localised nature of demand for housing and development generally, and Barker concludes that housing supply in areas of high demand is likely to exhibit far less responsiveness to price than other consumer durable products, such as cars, televisions and washing machines (Barker 2003: 42).

Key Points 5.3

- ❏ If only price changes, we move along the supply curve and there is an extension or contraction of supply.

- ❏ If any of the non-price determinants changes, the entire supply curve shifts to the left or right and we refer to a decrease or increase in supply (see Figure 5.2).

- ❏ Price elasticity of supply is given by the percentage change in quantity supplied divided by the percentage change in price.

- ❏ Long-run supply curves are more elastic than short-run supply curves because over a longer period of time more resources can flow into or out of an industry when price changes. This is especially the case in property markets.

- ❏ It is the inelastic supply of buildings that causes the related markets to be unstable and characterised by periods of escalating prices.

COMBINING SUPPLY AND DEMAND

In these chapters on demand and supply, we have confined our discussion to isolated parts of the market relating to the consumer or producer. Obviously this separation is theoretical and only useful for educational purposes. In reality, there is a very close relationship between the forces of demand and supply. Indeed, we have already discussed in Chapter 3 how the interaction of supply and demand determines prices. We introduced the concept of an equilibrium (or market) price at which both consumers' and producers' wishes are met. We even extended our discussion to consider the effects of changes to market conditions. Knowing and understanding how supply and demand interact is an essential prerequisite for interpreting many markets, including construction-related markets. It may be worthwhile, therefore, to spend a couple of minutes reviewing the market mechanism by attempting the exercise set out in the next section. An outline answer is provided in Figure 5.4 (shown on page 84).

EXERCISE ON SUPPLY AND DEMAND

Draw a basic supply and demand diagram (on a piece of paper) and consider the new equilibrium point in each of the three changing markets described in Table 5.3.

Table 5.3 Changing market conditions

Market	Product	Change in conditions	New equilibrium point
A	Energy efficient buildings	A successful advertising campaign by the energy efficiency office	See coordinate A in Figure 5.4
B	Computer assisted design	Improved microchip reduces cost	See coordinate B in Figure 5.4
C	New houses	A fall in the cost of a mortgage and an increase in the price of land	See coordinate C in Figure 5.4

Obviously this should be done before looking at the answers suggested overleaf (on page 84) in Figure 5.4.

Key Points 5.4

❑ By combining the forces of supply and demand we can begin to understand many markets.

❑ The concept of equilibrium demonstrates how the wishes of sellers and buyers are brought together via price.

❑ In reality, the determinants of supply and demand need to be considered simultaneously.

Figure 5.4 Changing market conditions across three markets

The three hypothetical markets and the changes described in Table 5.3 are plotted.
In market A we assume an increase in demand causing quantity and price to increase.
In market B we assume an increase in supply causing quantity to increase and price to
decrease. In market C we assume an increase in demand and a long-run decrease in
supply causing a higher equilibrium price but no significant short-run change to quantity.

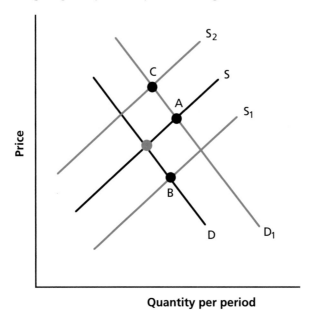

6 Clients and Contractors

In Chapters 4 and 5 we discussed two central ideas in explaining the market: demand and supply (see the Key Points in these chapters). In terms of construction economics, it is important to appreciate that those making the demands are referred to as clients and those who respond to their instructions by supplying the products are referred to as contractors. Both groups can take many forms, and it is common to emphasise that each project is a one-off. The analogy of a film set is sometimes employed to illustrate how the various contractors move in, do their work (complete a project) and then leave for another location to do 'similar' work for another client. Hence the industry is characterised by a fragmented approach, with each sub-group concentrating on its own specialism.

Extending an idea of Ive and Gruneberg (2000: 151), Table 6.1 tries to represent the innumerable projects that contractors could meet in response to their clients' demands. The location of a particular project, its design and production are some of the reasons why each project is unique. In the table, we also emphasise the fact that the construction stage is carried out by a vast combination of firms selected from the thousands comprising the industry. On each project there will be various teams of labour on site, working with different sets of people. This is in stark comparison to manufacturing and service industries where the work is usually repetitive and the workforce is normally permanent.

Table 6.1 Contractors involved in construction

	Project Types				
	BUILDING	CIVIL ENGINEERING	REFURBISHMENT	REPAIRS AND MAINTENANCE	
	A B C D	E F G H	I J K L	M N O P	etc.
DESIGN					
CONSTRUCTION	186,000 FIRMS				
OWNERSHIP					
USE					

Source: Adapted from Ive & Gruneberg (2000) and Construction Statistics Annual (BERR 2007: Table 3.1)

The perceptive reader may question the idea that all registered construction firms are only involved in the construction stage, as closer links are evolving between clients and contractors. These links will be discussed later in this chapter, under the heading of partnering. But first we need to describe the traditional arrangement of client and contractors as two separate parties involved in the construction of one project.

CLIENTS

At the core of any construction process are the clients. Some are well informed and know precisely what they want and how it can be technically achieved, but the majority seem to know nothing. Within the discipline of economics, a whole literature has begun to emerge that discusses transactions in which some of the parties involved know more than others. (This area of study is referred to as the economics of **asymmetric information** and will be considered further in Chapter 10 when we discuss market failure.) It has been suggested that independent advisers may be required to help clients. With a similar role to financial advisers, they would assist inexperienced clients to decide what is specified and how it can be best achieved. The case for client support has become stronger in recent years. As suggested in Chapter 2, the government can no longer act as a monolithic client. In fact, the changes in public sector emphasis have quite possibly fragmented the client base even further. As the Latham Report (1994: 3) was quick to point out: 'Nowadays the National Joint Consultative Committee for Buildings (NJCC) *has no means of ensuring* that all housing associations, trust hospitals, grant maintained schools, private government agencies, utilities companies, etc. are aware of the best current practice and changes in the construction industry.'

To address these problems the Construction Task Force led by Sir John Egan made comparisons with other industries that have increased efficiency as market forces expanded. The task force recognised that, in the best companies, the customer drives everything – the customer is 'king'. It is clear that the modern client in markets such as cars, steel and engineering products expects value for money, products that are free from defect, goods delivered on time, worthwhile guarantees and reasonable running costs. Unfortunately, the picture painted of construction in the late twentieth century was of an industry that 'tends not to think about the customer (either the client or the consumer), but more about the next employer in the contractual chain' (Egan 1998: 16).

CONTRACTORS

One of the most striking features of the construction industry across the world is the large number of firms. In the United Kingdom alone, there are approximately 186,000 private construction contractors recorded in the official government statistics and 93 per cent of them employ fewer than eight people. Across the European Union, there are more than 2.5 million contractors – employing 15 million people. Construction accounts for nearly 10 per cent of the total employment in

Europe. Much of this labour works in small enterprises. Small firms dominate the industry for two main reasons:

- they can supply services which do not suit the nature of large firms, such as repair and maintenance
- they can supply labour on a subcontract basis to large firms.

Labour-only Subcontractors

It is usual in Europe, and particularly in the UK, for the majority of work carried out on a construction site to be done by subcontractors. In addition, subcontractors often organise the materials and maintain the equipment. To the main contractor, labour-only subcontracting is the cheap and efficient option, as the self-employed worker is not entitled to holiday pay, redundancy money, sick leave, pension rights or any other benefits that accrue to permanent members of staff.

As a result of this tradition, a high level of fragmentation is often associated with the European construction industry. After being in post for six months a UK construction minister once remarked in an interview that in terms of meeting all the trade associations he was only half way down the alphabet. The interviewer sarcastically added that 'by the time he gets to Z he'll probably have moved on to his next job' (Broughton 2001). This characteristic of fragmentation leads to many of the industry's recognised strengths and weaknesses.

WEAKNESSES

There is a recognised lack of collaboration within construction teams. This in turn often leads to a lack of trust between the various parties and explains the adversarial nature that typifies the relationships. Another weakness of fragmentation is the lack of commitment to education, training, research and safety on site. It is often suggested that subcontractors do not seem to learn from one project to the next – a common myth is that each building is regarded as a 'prototype'. The general level of tension between contractors and subcontractors was interestingly captured when a recession began to loom at the start of 2008, as the first thing that the large contractors chose to do to protect their business was to streamline their supply chains and cut back on the number of specialist subcontractors employed. It was reported that Wates aimed to cut its suppliers list from 4,000 to 500 and Costains aimed to be equally as drastic with a cut from 18,000 to 2,500. To justify their actions the directors of these large organisations used derogatory phrases such as 'we will drop "white van man" in favour of bigger companies', 'we will weed out the small "fly-by-night" organisations as we seek to work with fewer quality subcontractors' and 'our aim is to be responsible for between 10 and 25 per cent of a subcontractor's work – [we want] to assure that our position (as the main contractor) is important to them, but not too important!' (McMeeken 2008). Furthermore, to compound the nature of this problematic relationship, specialists fortunate enough to remain on the main contractor lists are increasingly expected to jump through all kinds of hoops relating to health and safety and sustainability, which often involves paying for associated accreditation schemes.

A generic outcome of these types of problems is the way it limits any attempts that the construction industry seeks to make in the long term to construct on time, on budget and to the quality expected in a world concerned with sustainability.

STRENGTHS

On the positive side, it is usual to identify one strength of fragmentation: it enables the industry to be sufficiently flexible to deal with the highly variable workloads that accompany the changing economic circumstances.

Clearly the weaknesses outweigh the strengths and this is why the various types of partnering agreements that are emerging are regarded as such promising developments. These are discussed in the next section.

Key Points 6.1

❑ Clients create demand for the construction process. Increasingly, these clients are based in the private sector.

❑ Contractors liaise with the client to supply the products, which are typically produced by subcontractors.

❑ The construction team is typified by fragmentation, which leads to weaknesses such as a lack of collaboration, trust, training, etc.

PARTNERING

Since the publication of the Banwell Report in 1964, various government committees and academics have been advocating partnering as an important way of improving the performance of the construction industry. For example, a National Audit Office report (NAO 2001: 58) concluded that 'partnering is the way to deal with the inherent problems of construction which is still widely regarded as a design to order industry'. A comprehensive review of the academic literature suggests a consensus that partnering leads to better cost control, better quality product, closer relationships, enhanced communication, continuous improvement, potential for innovation, lower administrative cost, reduced litigation, increased satisfaction and improved culture (Chan et al. 2003: 524).

There are many definitions of **partnering** and different forms have been used in the UK, US and Japan. In general terms, however, partnering refers to some form of collaborative approach in which clients and contractors are increasingly open with one another in order to meet common objectives.

The idea of partnering began to gain momentum through the 1990s and there were a number of government reports and academic papers supporting its use. For example, Bennett and Jayes (1995) argued that partnering delivers cost savings ranging from as low as two per cent to as much as 30 per cent depending on the type of partnership. They found that **project partnering**, which is based on a client and contractor working openly on single project, produced cost savings of around 2–10 per cent. **Strategic partnering**, which involves the client and contractor working on a series of construction projects, was regarded as capable of delivering savings of up to 30 per cent. In addition, Bennett and Jayes (1995) specifically claim that partnering delivers better designs, makes construction safer, enables deadlines to be met with ease and provides all parties with increased profits. Some current examples of partnering will be dealt with in the following two sections to see how these possibilities can develop.

Prime Contracting

Prime contracting is a good example of partnering since it involves the integration of design, construction and maintenance under the leadership of one main contractor. The prime contractor takes on responsibility for everything from the selection of the subcontractors through to the delivery of the product. Consequently the client has only one point of contact, and the prime contractor strives to look after the client's interests and assure that quality, budgets and delivery targets improve.

Prime contractors are presently looking after estates run by the Ministry of Defence and new supermarkets being built on behalf of J. Sainsbury plc. Such contracts are usually continued for a number of years – on a repeat business basis. Accordingly, both parties seem to gain a greater opportunity to learn what is needed.

Private Finance Initiative (PFI)

This form of partnering has a far higher profile, since it involves the public and private sectors collaborating over some of the biggest projects presently being constructed. These projects include highways, hospitals, prisons, schools and government offices. Since its launch in November 1992, more than 650 PFI contracts have come into force representing future commitments of around £200 billion (HM Treasury 2007). Similar arrangements, such as public private partnerships, are a growing phenomenon worldwide.

A project procured under the **private finance initiative** is based on a distinctive kind of relationship between a public sector client and the private sector. The general procedure is as follows: private firms operating in a consortium agree to design, build, *finance* and *manage* a facility traditionally provided by the public sector. (Finance and manage are italicised to emphasise that a distinguishing characteristic of these schemes is that under PFI arrangements the private sector is expected to raise the initial finance to fund the project and thereafter manage its operation.) In return, the public sector client agrees to pay annual charges (called a unitary charge) during the life of the contract and/or allows the private sector to reap any profits that can be made for a specified period. The contracts usually run for 25 to 30 years or more. In this way, both sectors can be seen to be specialising in what they do best: with the public sector client initiating the development, specifying the requirements and vetting the tenders, and the private sector taking the risks and determining the best way to deliver the service.

The group formally awarded with the contract is referred to as a consortium or special purpose vehicle (SPV). The SPV may comprise several companies such as a construction and facilities management contractor, a bank and/or an equity investor, a management company, and separate liaising companies. (Subsequently, the SPV may subcontract its responsibilities.) This complex set of relationships is implied in Figure 6.1, where the PFI relationships have been simplified into a flow diagram.

Figure 6.1 Private finance initiative

PFI arrangements have the obvious advantage that the contractor no longer 'builds and disappears' – as in the BAD old days – because the *finance* and *manage* elements of the contract legally tie the contractor to the project once it has been built. When the private sector has money at risk in this way, there are far greater incentives to get everything right, especially as the consortium contractors now have to consider the running and maintenance costs.

In effect, PFI procurement creates a far better level of communication between public sector clients and private sector contractors, as both sides are linked in a common objective. The collaborative agreement provides mutual benefits to both parties. In the terms introduced in Chapter 2, the completed project should be more economically efficient and more sustainable (see Key Points 2.4 and see Figure 6.1).

Using private capital and expertise in the provision of public sector services is not entirely new. For example, joint working between sectors has a reasonably long history in housing – local authorities have transferred the ownership and management of large estates to a government regulated private sector group of **registered social landlords**. The point is that apart from the much debated private finance initiative, there are other forms of public private partnerships, where the public, private and voluntary sector work together to achieve a wide variety of objectives.

CONCLUSION

In conclusion, it is easy to see why partnering in any form is an attractive idea. In purely economic terms it clearly helps to eliminate inefficiency in so far as costs per unit of output are reduced. Leibenstein's (1973) concept of **x-inefficiency** can be applied. He argued that when a government department or individual firm is left entirely to its own devices to choose the APQT – that is, the activity, pace, quality of work, and time spent – it is unlikely to choose a combination that will maximise efficiency. In other words, if a public sector authority or individual contractor can choose what to build, how to build it, how long to spend on the project and so on, it is less likely to maximise the value of output. The resulting loss of value is what Leibenstein referred to as x-inefficiency and his solution relies on securing a more competitive market structure. Partnering potentially improves the dynamics of a market by putting the client more firmly in the driving seat, improving the flow of information between the participants and providing greater incentives to complete the contract on time, to budget and to the expected quality. Some of the benefits of partnering listed by the National Audit Office in its report on how the construction industry in the UK could become more efficient are summarised in Table 6.2.

Table 6.2 The benefits of partnering

✓	Reduced need for costly design changes
✓	Increased opportunities to replicate good practice learned on previous projects
✓	Avoids adversarial relationship between client and contractors
✓	Contractors have good incentives to deliver on time, on budget and to a high standard
✓	The liaison of clients and contractors should improve the overall efficiency of the building – particularly in terms of its operation and maintenance
✓	It should be possible to drive out inefficiency and waste from the construction process

Source: Adapted from National Audit Office (2001: 6, 31)

There are, however, some reservations about whether the benefits of partnering are fully realised under PFI. For instance, changes are inevitably required during the life of a contract running over 25 to 30 years, either as result of changing client needs or because of new technological developments, and these can prove to be particularly expensive to implement, especially when one company has already been nominated as the main contractor. The National Audit Office (2008) has confirmed that changes to PFI contracts tend to be more expensive than equivalent changes in conventionally procured projects by a margin of 5 to 10 per cent of the cost of the change. When it is noted that public authorities spent over £180 million in 2006 on changes to operational PFI projects, the scale of the problem becomes evident.

A second issue is that the annual (unitary) costs are increasing in size as PFI extends. The health service alone paid £470 million in the year 2005–06 to cover the 80 schemes that were then running in England; by 2013–14 the equivalent bill covering 126 schemes is forecast to cost £2.3 billion! This could create financial difficulties in managing the budgets allocated for health, and it is argued that this could lead to service cuts and closures (Hellowell and Pollock 2007).

Another significant drawback is that successful partnering is far easier for big firms than small firms – and construction is still dominated by the latter. In a truly competitive perfect market all information is freely available – everybody has access to the knowledge they need for exchange to take place – and transaction costs are zero. Governments often try to ensure these conditions prevail by standardising legal and financial procedures, in order that market participants know where they stand. However, in construction markets where partnering and, in particular, PFI is emerging these conditions do not prevail. Many firms lack the necessary resources to understand the complex legal information that is inevitably associated with these forms of procurement. Transaction costs are prohibitively high, with architects', lawyers' and accountants' fees to be met by all the participating parties. As a result, it is unusual for more than three or four consortium groups to find sufficient resources to engage in the tedious, lengthy and detailed bidding processes involved. Indeed, PFI bidding costs can commonly exceed £1 million per project. The firms that are able to take on such large-scale operations and risks are few and far between, and it is a common concern that partnering arrangements often exclude smaller contractors.

To compound these issues even further a recent directive from the European Commission suggests that at least five consortia need to compete in the final bid for major public projects. The National Audit Office and the Office of Fair Trading in the UK do not go quite as far, but they strongly recommend that competition for public sector procurement should involve at least three competing tenders. These developments could present yet another hurdle in developing and extending PFI and prime contracting arrangements.

Key Points 6.2

❏ Several government reports, backed by considerable academic writing, have advocated partnering as a vital part of improving efficiency in the construction industry.

❏ Partnering is a broadly used term describing several different types of collaborative arrangement.

❏ Partnering has made great strides in recent years – particularly in the public sector.

❏ Partnering assumes a win-win scenario for all parties. Several of the benefits are summarised in Table 6.2

❏ It is easier for large firms to gain advantage and benefit from partnering arrangements. Small firms generally lack the resources and expertise to participate in the lengthy bidding process involved in PFI contracts.

RETHINKING CLIENT AND CONTRACTOR RELATIONSHIPS

From the commentary in this chapter, it appears that the construction industry may be characterised by dissatisfaction and, compared to other industries, may seem inefficient. Indeed, Sir John Egan alluded to the severe problems facing the industry in the late 1990s when he commented on the low and unreliable rate of profits that prevent the necessary investment in training, research and development to sustain a healthy industry. In fact, he noted that the City's view of construction confirmed its position as a poor investment. As Egan (1998: 11) observed: 'The City regards construction as a business that is unpredictable, competitive only on price not quality, with too few barriers to entry for poor performers.' From the client's perspective investment was portrayed in a similar way, as unpredictable in terms of delivery, budget and the standards of quality expected. Again in Egan's (1998: 10) words: 'Investment in construction is seen as expensive, when compared both to other goods and services and to other countries. In short, construction too often fails to meet the needs of modern business that must be competitive in international markets, and rarely provides best value for clients and taxpayers.'

A subsequent review of government R&D policies for the construction industry, *Rethinking Construction Innovation and Research* (Fairclough 2002), further confirmed the impression of underachievement. Indeed, Fairclough portrayed an industry at the beginning of the twenty-first century that was 'dirty, dangerous and old fashioned' (2002: 30). He emphasised that for construction to evolve a strategic vision was needed. Interestingly, the vision that was articulated centred on the industry's contribution to the overall aims of economic, environmental and social sustainability.

To paraphrase three of Fairclough's findings:

- the construction industry has a profound influence over the quality of our lives at home and at work (2002: 6) and has a key role to play in society in providing a better built environment (2002: 30)
- the construction sector suffers from a lack of focus and is unable to speak with a single voice on those issues that influence it, and this represents a significant challenge to the entire construction community – both clients and contractors – which must demand buildings whose economics are considered on a whole-life basis (2002: 6)
- the new vision for construction and its contribution to the wider quality of life should draw ideas from other industries, to improve its productivity and competitiveness and understand the relationship of these economic issues to sustainability (2002: 31).

The findings of Egan (1998) and Fairclough (2002) were drafted at the request of the UK government, and they captured the state of play at the beginning of the twenty-first century. Although the image presented was in many ways more 'Victorian' than 'modern', it applied with equal measure to most other nations. As evidence there is an international plethora of government-sponsored reports and in-depth reviews into the construction industry. For example, to paraphrase the opening paragraph of a review of the state of construction in Singapore, the construction industry has a poor image and it can easily be singled out from the rest of the economy by attitudes, technologies, processes and culture that are at least half a century old (Dulaimi et al. 2001: 1). Similar sentiments are expressed in the conclusion to Manseau and Seaden's (2001) international review of public policies in Europe, North and South America, South Africa and Japan, where they conclude that 'construction almost everywhere is perceived as being "in trouble", with low margins of profit, high costs of production and lack of concern for the end user'. Such observations confirm that construction has a key role to play in the progress of a sustainable society, but first of all the level of efficiency and communication needs to be improved between clients and contractors.

To some extent these difficulties explain the current emphasis that is being placed on **value management**. Although the process originated over 60 years ago within the US manufacturing sector and was referred to as value engineering, it has been increasingly applied to large-scale construction projects since the 1980s. For our purposes the terms value management and value engineering are synonymous, but in the UK the former term seems to be used more frequently.

VALUE MANAGEMENT

In the most general sense, value management is a specific event organised to identify and eliminate unnecessary costs, although it should not be interpreted as solely a cost-cutting exercise. This might sound confusing but the aim is to optimise value for money by reviewing costs, quality and function. There is no guarantee, therefore, that the initial costs will be cut, but a more efficient project should almost certainly emerge.

The process developed in the industrial community of the United States and an early anecdote outlines its origins. The story goes that Lawrence Miles a purchasing engineer of the General Electric Company during the Second World War was assigned the task of procuring materials to expand the production of the B24 bomber from 50 to 1000 aircraft per week. At the time there was a shortage of materials and consequently he initially had problems sourcing supplies. However, Miles discovered many substitute components that were not only cheaper but better than the original products. Hence the idea of value management was born and was associated with cost cutting. Today, however, it has broadened in scope and is far more closely associated with securing value for money, in the sense of optimising the whole-life costs of a project to meet the client's long-term requirements (HM Treasury 2006: 7).

The usual way that value management is implemented is through structured workshops, led by an independent facilitator, to take account of issues relating to design, construction, operation and management. In short, value management workshops are multi-disciplined. The aim is to review the performance and costs of a project throughout its life cycle. The prototype value management workshops were organised to take place over several days (often up to one week), but recently the trend has been for these events to be compressed into a far shorter time frame. In fact, research evidence suggests that half-day sessions are becoming the norm, and in certain circumstances this is even reduced to a couple of hours. To paraphrase the comments of a facilitator explaining this trend: the average client has a horror of spending more than £5000 on me [as facilitator] and has a cynical view of the value of 16 people sitting round a table discussing the project (Ellis et al. 2005: 489).

The precise format and timing of workshops will vary from project to project. In theory, there is nothing stopping a value management review from taking place at any stage of the process. Indeed, the theoretical literature even recommends that a kind of post-occupancy review should be carried out to document the lessons learnt and provide feedback. In practice, however, it has been found that the process rarely proceeds beyond the early stages, as although the value of post-completion exercises is accepted, these are rarely carried out (Ellis et al. 2005: 486).

In Figure 6.2 we outline a typical project life cycle to emphasise that the earlier a value management study is undertaken the more opportunities arise to benefit from it. This is especially important as in the majority of cases there is only one value management intervention (Ellis et al. 2005: 486). So if value management is to be used to good effect, most participants agree that they need to become involved as early as possible in the project's cycle when clear choices have to be made. For example, by holding a workshop to examine the outline business case, allows time and discussion for all stakeholders to gain understanding of the project, improve its efficiency and overall value for money. Providing the value management event is held before the tender stage, there is still opportunity to discuss different options, seek better solutions, accommodate design changes and generally eliminate unnecessary costs. Ideally this would then be followed by a series of value management events as the project unfolds, and there are exceptional cases where several value management events are spread across a number of years.

Value management is clearly an organised effort to analyse functions, value, costs and sustainability. In short, it looks at construction projects as an integrated whole. It is not surprising that the Office of Government Commerce (OGC), which seek to meet business requirements and deliver successful projects, encourages value management reviews on all large complex projects. Where the value management technique has been applied, the OGC boasts of a significant reduction in defects, effective redesign of key operational areas, reductions in maintenance charges and overall savings averaging 15 per cent (OGC 2007).

Figure 6.2 Project life cycle

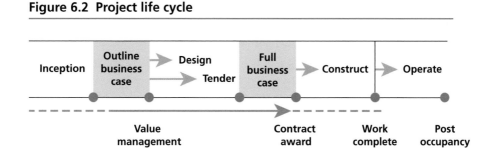

The United States Federal Highway Administration is even further committed to value management and has made it compulsory for projects exceeding $25 million (£13.8 million). As a consequence, it can refer to some impressive state-of-the-art examples. To take just one instance, in 2004 a project began to replace the Antler's bridge crossing across the Sacramento River arm of Lake Shasta with a new £58 million structure. A total of eight value management studies were undertaken – at the design stage, at the preliminary engineering stage, and so on. These resulted in an all-embracing value management process far ahead of anything currently carried out by the UK government. One major distinction was that this particular value management exercise included an assessment of the non-monetary benefits, such as travelling time and reducing the impact on nesting eagles, alongside the assessment of conventional monetary benefits such as construction and maintenance costs. An assessment of the revised design for the Antler's bridge crossing (following value management) showed a 19 per cent overall performance improvement. This was achieved at a reduction in cost of £7 million or 12 per cent. According to a formula used by the Californian government (dividing benefits by savings), these figures overall represented a 35 per cent improvement in value for money (OGC 2007).

Finally, in the context of this chapter, it should not be forgotten that value management equally provides an excellent opportunity to build a competent team and achieve effective collaboration between clients and contractors.

Key Points 6.3

❑ Compared to other industries, the processes associated with the traditional construction industry appear old fashioned, inefficient and unsustainable.

❑ The Egan Report (1998) emphasised that the industry needs to modernise.

❑ The Fairclough Report (2002) reviewed government R&D policies and confirmed that the construction industry would continue to underachieve until it adopted a coherent and sustainable vision.

❑ Value management is not merely a cost-cutting exercise, carried out properly it should address the relationships between cost, function and value.

❑ Value management is increasingly used to assess value for money on large-scale complex projects.

7 Costs of the Construction Firm

We now begin to examine more closely the supply side of an economy – that is, we develop a theory of how suppliers behave. This part of economics is referred to as the theory of the firm. How do business owners react to changing taxes, changing input prices, changing government regulations and changing market conditions? In order to answer these questions, we have to understand the nature of production costs and revenues for each firm. In this chapter, we examine closely the nature of productivity and costs. These costs may relate to design, construction, maintenance, management, conservation, refurbishment, or whatever. In Chapter 8, we bring in the revenue side of the picture and consider the level of profits typically available to construction firms. Before commencing this chapter, therefore, it may be useful to remind yourself of some basic ideas by reviewing the Key Points 1.2, 2.4 and 5.1.

THE FIRM

We start by defining a business, or firm, in general terms.

> A firm is an organisation that brings together different factors of production, such as labour, land and capital, to produce a product or service which it is hoped can be sold for a profit.

The actual size of a firm will affect its precise structure. A common set-up for a larger firm involves entrepreneur, managers and workers. The entrepreneur is the person who takes the chances. Because of this, the entrepreneur is the one who will get any profits that are made. The entrepreneur also decides who runs the firm. Some economists maintain that the true quality of an entrepreneur is the ability to pick good managers. Managers, in turn, are the ones who decide who should be hired and fired and how the business should be generally organised. The workers are the people who ultimately use the machines to produce the products or services that are being sold by the firm. Workers and managers are paid contractual wages. They receive a specified amount for a given time-period. Entrepreneurs are not paid contractual wages. They receive no specified 'reward'. Rather, they receive what is left over, if anything, after all expenses have been paid. Profits are, therefore, the reward to the entrepreneur for taking risks. Note that roles can be combined, and in the many small firms that comprise a large part of the construction industry the entrepreneur is also the manager-proprietor.

According to EU data, small and medium-sized enterprises – defined as firms employing less than 250 people – produce approximately 99 per cent of the construction output in Europe. In fact, more than 20 per cent of the European construction sector workforce is estimated to be self-employed; the comparable figure is closer to 40 per cent in the case of the UK. Indeed, the construction industry may be characterised by risk-taking entrepreneurs.

Profit

The costs of production must include an element of profit to pay for the entrepreneur's services. If the level of profits falls in one area of activity, entrepreneurs may move their resources to an industry where the returns are higher. To illustrate this behaviour economists employ a concept of **normal profit**. Normal profit may be defined as:

> the minimum level of reward required to ensure that existing entrepreneurs are prepared to remain in their present area of production.

Normal profit is included in the cost of production, as it is an essential minimum reward necessary to attract the entrepreneur into economic activity. The concept of normal profit also highlights that all resources can be employed in several ways (that is, all resources have alternative uses). Note that what is meant by 'profit' by economists differs from its general meaning in everyday usage. To portray the general everyday meaning of profit, the following formula could be used:

> profits = total revenues – total costs

For economists, an alternative formula is required:

> economic profits = total revenues – total opportunity cost of all inputs used

The economic profits formula will become clearer by looking at two areas of resource allocation and the related cost accounting calculations. The first resource is capital, and the second is labour.

OPPORTUNITY COST OF CAPITAL

Firms enter or remain in an industry if they earn, at a minimum, a normal rate of return (NROR) – that is, normal profit. By this term, we mean that people will not invest their wealth in a business unless they obtain a positive competitive rate of return – in other words, unless their invested wealth pays off. Any business wishing to attract capital must expect to pay at least the same rate of return on the capital as all other businesses of *similar risk* are willing to pay. For example, if individuals can invest their wealth in almost any construction firm and get a return of 10 per cent per year, then each firm in the construction industry must expect to pay 10 per cent as the normal rate of return to present and future investors. This 10 per cent is a cost to the firm – formally referred to as the **opportunity cost of capital**. The opportunity cost of capital is the amount of income, or yield, forgone by giving up an investment in another firm. Capital will not stay in firms or industries if the rate of return falls below its opportunity cost. Clearly, the expected rate of return will differ from industry to industry according to the degree of risk and difficulty involved.

OPPORTUNITY COST OF LABOUR

The self-employed contractor or one-person business often grossly exaggerates profit, because the opportunity cost of the time that they personally spend in the business is often not properly measured. Take, as an example, the people who run small surveying offices. These surveyors, at the end of the year, will sit down and

figure out their 'profits'. They will add up all their fees and subtract what they had
to pay to staff, what they had to pay to their suppliers, what they had to pay in
taxes, and so on. The end result, they will call 'profit'. However, they will not have
figured into their costs the salary that they could have earned if they had worked for
another company in a similar type of job. For a surveyor, that salary might be equal
to, say, £20 an hour. If so, then £20 an hour is the opportunity cost of the surveyor's
time. In many cases, people who run their own businesses lose money in an
economic sense. That is, their profits, as they calculate them, may be less than the
amount they *could* have earned had they spent the same time working for someone
else. Take a numerical example. If an entrepreneur can earn £20 per hour, it follows
that the opportunity cost of his or her time is £20 x 40 hours x 52 weeks, or
£41,600 per year. If this entrepreneur is making less than £41,600 per year in
accounting profits, he or she is actually losing money. This does not mean that such
entrepreneurs are stupid: they may be willing to pay for the non-pecuniary benefits
of being the boss.

We have only considered the opportunity cost of capital and the opportunity
cost of labour, but the concept applies to all inputs. Whatever the input, its
opportunity cost must be taken into account when figuring out true economic
profits. Another way of looking at the opportunity cost of running a business is that
opportunity cost consists of all explicit (direct) and implicit (indirect) costs.
Accountants only take account of explicit costs. Therefore, accounting profit ends
up being the residual after only explicit costs are subtracted from total revenues.

Figure 7.1 Simplified view of economic and accounting profit

Here we see that on the left-hand side, total revenues are equal to accounting cost
plus accounting profit – that is, accounting profit is the difference between total
revenue and total accounting cost. On the other hand, we see in the right-hand column
that economic profit is equal to total revenue minus economic cost. Economic costs
equal explicit accounting costs plus a normal rate of return on invested capital (plus any
other implicit costs).

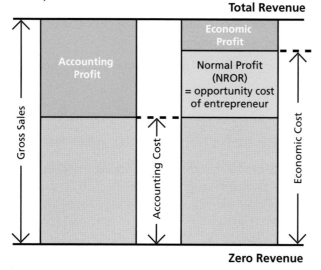

The term profits in economics means the income that entrepreneurs earn over and above their own opportunity cost of time, plus the opportunity cost of the capital they have invested in their business. Profits can be regarded as total revenues minus total costs – which is how the accountants think of them – but economists include *all* costs. We indicate this relationship in Figure 7.1.

The Goal of the Firm

In developing a model of the firm, we will generally assume that main business goal is maximisation of profits. In other words, a firm's goal is to make the positive difference between total revenues and total cost as large as it can. We use this **profit-maximising model** because it allows us to analyse a firm's behaviour with respect to the relationship between costs and units of output. Whenever this model produces poor predictions, we will examine our initial assumption about profit maximisation. We might have to conclude that the primary goal of *some* firms is not to maximise profits but rather to maximise sales, or the number of workers, or the prestige of the owners, and so on. However, we are primarily concerned with generalisations. Therefore, providing the assumption of profit maximisation is correct for *most* firms, then the model serves as a good starting point.

Key Points 7.1

❑ A firm is any organisation that brings together production inputs in order to produce a good or service that can be sold for a profit.

❑ Accounting profits differ from economic profits.

❑ Economic profits are defined as total revenues minus total costs, including the full opportunity cost of all the factors of production.

❑ Small businesses and self-employed workers often fail to consider the opportunity cost of the labour services provided by the owner.

❑ The full opportunity cost of capital invested in a business is generally not included as a cost when accounting profits are calculated. Thus, accounting profits overstate economic profits.

❑ Profit maximisation is regarded as the main objective when considering a firm's behaviour.

THE RELATIONSHIP BETWEEN OUTPUT AND INPUTS

A firm takes numerous inputs, combines them using a technological production process, and ends up with an output. There are, of course, many factors of production, or inputs. We can classify production inputs into two broad categories – labour and capital. The relationship between output and these two inputs is as follows:

output per unit of time = some function of capital and labour inputs

For more than 25 years, Patricia Hillebrandt (1974: 34; 2000: 191) used this type of framework to discuss the output of the whole construction industry. In her terms, the ultimate determinants of capacity within the industry are as follows:

- the amount of a given resource in use to sustain the current level of output
- the amount of the resource lying idle and awaiting the demand for its use
- the extent to which a resource may increase during the time horizon under review
- the potential to substitute one resource (such as labour) for another (such as capital) – and, obviously, the related cost of substitution.

Short Run Versus Long Run

Clearly, in any discussion of capacity, the time period is important and of particular significance. Throughout the rest of this chapter, we will consider a 'short' time period as opposed to a 'long' time period. In other words, in the analysis that follows we are looking at *short-run* production relationships and *short-run* costs associated with production.

Any definition of the short run is, necessarily, arbitrary. We cannot talk in terms of the short run being a specific period such as a month, or even a year. Rather, we must consider short run in terms of the ability of the firm to alter the quantity of its inputs. For ease of understanding, we will simply define the **short run** as any time period when there is at least one factor of production that has a fixed cost. In the **long run**, therefore, all costs are variable; that is, all factors are variable.

How long is the long run? That depends on the individual industry. For retailing, the long run may be four or five months – because that is the kind of time period in which they can add new franchises. For manufacturing firms, the long run may be several years – because that is how long it takes to plan and build a new factory.

In most short-run analyses, the factor that has a fixed cost, or is fixed in quantity, is capital. We therefore state that in our short-run model, capital (as well as land) is fixed and invariable. This is not unreasonable – on a typical construction site the amount of land and the number of cranes in place will not change over several months. Likewise, the head office location and equipment can not be easily altered. The input that changes most is labour. The production relationship that economists use, therefore, holds capital and land constant, or given, and assumes labour is variable. This assumption is particularly pertinent in an industry such as construction in which labour accounts for a significant proportion of the costs.

The Production Function – a Numerical Example

The relationship between physical output and the quantity of capital and labour used in the production process is sometimes called a **production function**. The term production function in economics owes its origin to production engineers for it is used to describe the technological relationship between inputs and outputs. It depends therefore on the available technology.

Look at Table 7.1 (on page 102). Here we show a production function that relates total output in column 2 to the quantity of labour input in column 1. When there are no workers, there is no output. When there is the input of five workers (given the capital stock), there is a total output of 150 sq. m per week. (Ignore, for the moment, the rest of that Table 7.1.) In Figure 7.2a, we show this hypothetical production function graphically. Remember that it relates to the short run and that it is for an individual firm.

Table 7.1 Diminishing returns: a hypothetical case in construction

In the first column, we give the number of workers used per week on a project. In the second column, we give their total product; that is, the output that each specified number of workers can produce in terms of square metres. The third column gives the marginal product. The marginal product is the difference between the output possible with a given number of workers minus the output made possible with one less worker. For example, the marginal product of a fourth worker is 20 square metres, because with four workers 140 square metres are produced, but with three workers only 120 are produced; the difference is twenty.

Input of labour	Total product (output in sq. m per week)	Marginal physical product (in sq. m per week)
0	0	
1	20	20
2	60	40
3	120	60
4	140	20
5	150	10
6	160	10
7	165	5
8	163	−2

Figure 7.2a shows a total physical product curve, or the amount of physical output that is possible when we add successive units of labour while holding all other inputs constant. Note that the graph of the production function in Figure 7.2a is not a straight line. In fact, it peaks at seven workers and starts to go down. Obviously the unique nature of each construction related project prevents us predicting an optimum number for each job, especially as the particular skills of the construction team also greatly influence the actual rate of output. In general terms, however, within any sector of the construction industry there is a recognisable trend: output increases as more units of labour are taken on but not at a constant rate. To understand why such a phenomenon occurs within any firm in the short run, we have to analyse in detail the law of diminishing (marginal) returns.

Figure 7.2a A production function

A production function relates outputs to inputs. We have merely taken the numbers from columns 1 and 2 of Table 7.1 and presented them as a graph.

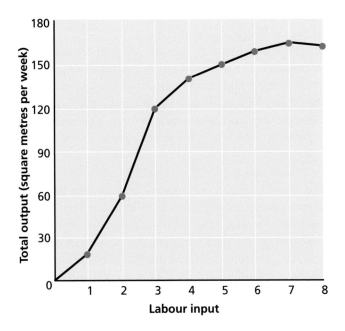

DIMINISHING RETURNS

The concept of diminishing marginal returns applies to many different situations. If you buckle a seat belt over your body in a car, a certain amount of additional safety is obtained. If you add another seat belt some more safety is obtained, but not as much as when the first belt was secured. When you add a third seat belt, again safety increases but the amount of *additional* safety obtained is even smaller. In a similar way, the **u-values** – a measure of heat loss – related to glazing do not decline steadily as you add more panes of glass within a window unit. The u-values typically associated with single, double and treble glazing are 5.7, 2.8 and 2.0 respectively. Therefore, assuming the wall construction and other factors remain constant, going from single to double glazing improves the u-value by 2.9, while adding a third pane of glass only improves the u-value by 0.8.

The same analysis holds for firms in their use of productive inputs. When the returns from hiring more workers are diminishing, it does not necessarily mean that more workers will not be hired. In fact, workers will be hired until the returns, in terms of the *value* of the extra output produced, are equal to the additional wages that have to be paid for those workers to produce the extra output. Before we get into the decision-making process, let us demonstrate that diminishing returns can easily be represented graphically and subsequently used in our analysis of the firm.

Measuring Diminishing Returns

How do we measure diminishing returns? First, we will limit the analysis to only one variable factor of production (or input). Let us say that factor is labour. Every other factor of production, such as machinery, *must* be held constant. Only in this way can we calculate the marginal returns from using more workers and know when we reach the point of diminishing marginal returns.

Marginal returns for productive inputs are sometimes referred to as the **marginal physical product**. The marginal physical product of a worker, for example, is the change in total product that occurs when that worker joins an already existing team. It is also the *change* in total product that occurs when a worker resigns or is laid off from an already existing construction project. The marginal productivity of labour, therefore, refers to the change in output caused by a one-unit change in the labour input.

At the very beginning, the marginal productivity of labour may increase. Take a firm starting without any workers, only machines. The firm then hires one worker, who finds it difficult to do the work alone. When the firm hires more workers, however, each is able to *specialise*, and the marginal productivity of these additional workers may actually be greater than that achieved with the few workers. Therefore, at the outset, increasing marginal returns are likely to be experienced. Beyond a certain point, however, diminishing returns must set in; each worker has (on average) fewer machines with which to work (remember, all other inputs are fixed). Eventually, the site will become so crowded that workers will start running into one another and will become less productive.

Using these ideas, we can define the **law of diminishing (marginal) returns**. Consider this definition.

As successive equal increases in a variable factor of production, such as labour, are added to other fixed factors of production, such as capital, there will be a point beyond which the extra or marginal product that can be attributed to each additional unit of the variable factor of production will decline.

We can express this more formally.

As the proportion of one factor in a combination of factors is increased, after a point, the marginal product of that factor will diminish.

Put simply, diminishing returns means that output does not rise in exact proportions to increases in inputs, such as the number of workers employed.

AN EXAMPLE

The law of diminishing returns can be demonstrated in any sector of the economy. Take a building site as an example. There is a fixed amount of land (the building plot), and with the necessary supply of materials (bricks, timber, etc.) and certain tools, the addition of more labourers eventually yields decreasing increases in output. A hypothetical set of numbers, based on discussions with project managers, that illustrates the law of diminishing returns has already been presented in Table 7.1 (and you may need to revisit the related caption). This is now represented graphically in Figure 7.2b.

Marginal productivity (the returns from adding more workers) at first increases, then decreases, and finally becomes negative. When one worker is hired, total output goes from zero to 20. Thus, the marginal physical product is equal to 20. When another worker is added, the total output, measured in square metres per week, increase to 60. Consequently the marginal physical product associated with the second unit of labour is 40. The third unit of labour adds a further 60 square metres to the total and, thereafter, the marginal product begins to decrease. In this example, therefore, diminishing marginal returns occurs after three workers are hired.

Figure 7.2b Diminishing marginal returns

Taking the data from Table 7.1, on the horizontal axis we plot the numbers of workers and on the vertical axis we plot the marginal physical product in square metres per week. When we go from zero workers to one worker, marginal product is 20. We show this at a point between 0 and 1 worker to indicate that the marginal product relates to a change in the total product as we add additional workers. When we go from one to two workers, the marginal product increases to 40. After three workers, marginal product declines. Therefore, after three workers, we are in the area of diminishing marginal physical returns. Total product, or output, reaches its peak at seven workers. In fact when we move from seven to eight workers, marginal product becomes negative.

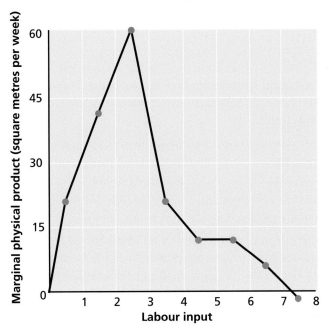

Diminishing Marginal Returns and the Theory of the Firm

If we now introduce business costs into the picture, we can begin to understand the central importance of the law of diminishing returns. For example, consider the relationship between marginal cost – that is, the cost of an extra unit of output – and the incidence of diminishing marginal physical returns as illustrated in

Table 7.1. Let us assume that each unit of labour can be purchased at a constant price. Further assume that labour is the only variable input. We see that as more workers are hired, the marginal physical product first rises and then falls after the point at which diminishing returns are encountered. The marginal cost of *each extra unit* of output will first fall as long as the marginal physical product is rising, and then it will rise as long as the marginal physical product is falling. Consider again the data in Table 7.1. Assume that a worker is paid £500 a week. When we go from zero labour input to one unit, output increases by 20 square metres. Thus, the labour cost per square metre (the marginal cost) is £25. Now the second unit of labour is hired, and it, too, costs £500. Output increases by 40. Thus, the marginal cost is £12.50 (£500 divided by 40). We continue the experiment. The next unit of labour produces 60 additional square metres, so the marginal cost falls further to £8.33 per square metre. The next labourer yields only 20 additional square metres, so the marginal costs starts to rise again back to £25. The following unit of labour increases marginal physical product by only 10, so that marginal cost becomes £50 per square metre (£500 divided by 10).

Marginal costs in turn affect the pattern of other costs, such as average variable costs and average total costs. Once these other costs have been considered, the importance of marginal cost analysis (and this whole section) will become clearer.

Key Points 7.2

❑ The technological relationship between output and input is called the production function. It relates output per unit of time to the inputs, such as capital and labour.

❑ After some rate of output, the firm generally experiences diminishing marginal returns.

❑ The law of diminishing returns states that if all but one of the factors of production are held constant, equal increments in that one variable factor will eventually yield decreasing increments in output.

❑ A firm's short-run costs are a reflection of the law of diminishing marginal returns. Given any constant price of the variable input, marginal costs decline as long as the marginal product of the variable resource goes up. At the point of diminishing marginal returns, the reverse occurs. Marginal costs will rise as the marginal product of the variable input declines.

SHORT-RUN COSTS

In the short run, a firm incurs different types of costs. Economists label all costs incurred as total costs. These are divided into fixed costs and variable costs, which we explain below. The relationship, or identity, is therefore:

total costs = total fixed costs + total variable costs

Examples of costs in each of these categories are introduced in Table 7.2. After we have looked at the elements of total costs, we will find out how to compute average and marginal costs.

Table 7.2 Typical construction costs

Type of Cost	Examples
Variable Costs	<ul><li>Labour used on site</li><li>Materials used on site</li><li>Equipment used on site</li><li>Site management</li><li>Tendering for future contracts</li></ul>
Fixed Costs	<ul><li>Head office bills for energy, water and rates</li><li>Wages for permanent head office staff</li><li>Bank interest and leasing costs</li><li>A sufficient (normal) level of return to keep the entrepreneur in the industry</li></ul>

Fixed Costs

Let us look at a building firm such as Taylor Wimpey. The managers of that business can look around and see the plant and machinery that Taylor Wimpey owns, the office buildings the company occupies and the permanent staff for whom they are responsible. Taylor Wimpey has to take account of the wear and tear of this equipment and pay the administrative staff, no matter how many houses it builds. In other words, all these costs are unaffected by variations in the amount of output. This leads us to a very straightforward definition of **fixed costs**.

All costs that do not vary – that is, costs that do not depend on the rate of production – are called fixed costs, or sunk costs.

According to Baumol's (1982) theory of contestable markets, when sunk costs are low, existing firms in a market have a continual concern that new entry is always possible. In these circumstances profit levels are constrained (see Chapter 8 for details). Contestable markets typify construction firms, as they usually have quite low fixed costs. Most contractors have no factory – as in effect each site represents the firm's new work location – and much of the necessary equipment, such as scaffolding, cranes, skips, toilets, office huts, security stores, floodlights and even water supplies, is usually hired as and when required.

Variable Costs

The difference between total costs and total fixed costs is total variable costs: that is, total costs – total fixed costs = total variable costs. **Variable costs** are those costs whose magnitude varies with the rate of production. As a proportion of total costs, variable costs in construction tend to be much higher than in the manufacturing industry. One obvious variable cost is wages. The more a firm builds or makes, the more labour it has to hire and the more wages it has to pay. In fact a limiting factor to a construction firm's output is often its management: unlike machines, managers do not have an automatic safety cut-off point when they are operating at full capacity, and when things start to go wrong on site, costs can quickly spiral. As the size and number of their projects increases, construction firms need to employ good site and project managers, and these may well be in short supply. It is, therefore, difficult to determine when variable costs will rise. The same type of logic applies to the other main variable cost category – materials. As the demand for materials increases, they too may well become more expensive to acquire.

In construction, the distinction between fixed and variable costs can be difficult to make. For example, some contractors may regard management salaries as fixed costs. It depends how the individual firm is organised. However, the way to avoid a high proportion of fixed costs is to meet any requirement to increase output by subcontracting and this is largely what happens in the construction industry across Europe.

Short-Run Average Cost Curves

In Figure 7.3a, we plot total costs, total variable costs and total fixed costs. You should note that the variable cost curve lies below the total cost curve by the vertical distance equivalent to total fixed costs. In manufacturing firms, the vertical distance representing fixed costs will be greater because its activities are based on a factory which probably houses expensive machinery and so it has relatively high fixed costs. Figure 7.3a is meant to represent the cost curves of a typical construction firm – the fixed costs are represented as proportionally low.

Next we want to look at the average cost. The average cost is simply the cost per unit of output. It is a matter of simple arithmetic to calculate the averages of these three cost concepts. We can define them simply as follows:

$$\text{average total costs} = \frac{\text{total costs}}{\text{output}}$$

$$\text{average variable costs} = \frac{\text{total variable costs}}{\text{output}}$$

$$\text{average fixed costs} = \frac{\text{total fixed costs}}{\text{output}}$$

Figure 7.3b shows these corresponding average costs. Let us see what we can observe about the three average cost curves in that graph.

Figure 7.3a Total costs of production

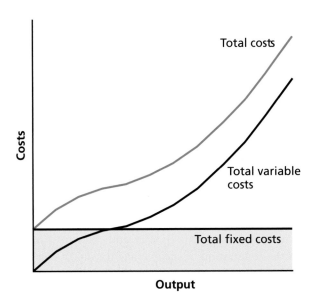

Figure 7.3b Average fixed costs, average variable costs
average total costs and the marginal costs of production

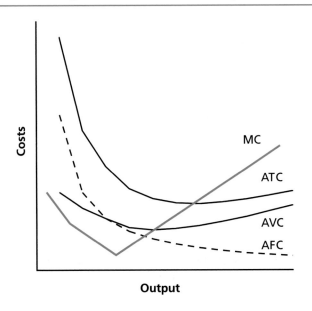

AVERAGE FIXED COSTS (AFC)

As we can see from Figure 7.3b, **average fixed costs** fall throughout the output range. In fact, if we were to continue the diagram further to the right, we would find that average fixed costs would get closer and closer to the horizontal axis. This is because total fixed costs remain constant. As we divide this fixed amount by a larger number of units of output, AFC must become smaller and smaller.

AVERAGE VARIABLE COSTS (AVC)

We assume a particular form of the **average variable cost** curve. The form that it takes is a flattened U-shape: first it falls; then it starts to rise. The shape of the curve indicates that at first it costs less to build or make successive units, but as the law of diminishing returns sets in, it costs more and more to make successive units.

AVERAGE TOTAL COSTS (ATC)

This curve has a shape similar to the average variable cost curve. However, it falls even more dramatically at the beginning of the output range and rises more slowly after it has reached a minimum point. It falls and rises in this fashion because **average total costs** is the summation of the average fixed cost curve and the average variable cost curve. So, when AFC plus AVC are both falling, it is only logical that ATC will fall too. At some point, however, AVC starts to rise while AFC continues to fall. Once the increase in the AVC curve outweighs the decrease in the AFC curve, the ATC curve will start to increase and it will develop its U-shape.

An efficient firm will aim to achieve its output at the lowest point on the average cost curve – as this is where each unit of production is associated with its lowest possible cost, given the firm's existing level of capacity.

Marginal Cost

To highlight precisely when average costs reach their lowest point, economists are very interested in the principle of **marginal cost**. As we discussed earlier in this chapter in the section on diminishing returns, the term marginal means additional or incremental. In our previous example set out in Table 7.1, we considered the marginal physical product – that is, the additional output associated with taking on successive units of labour – and the marginal costs of that additional output. For convenience the data of Table 7.1 is reproduced in Table 7.3 together with columns setting out the marginal costs and average variable costs. These are calculated using the same assumptions – that each unit of labour costs £500 per week and that these are the only costs that alter – and the same method as on page 106. For example, when the second unit of labour is hired, costing £500, output increases by 40. Thus, the marginal cost is £12.50 (£500 ÷ 40) per square metre

Remember we find marginal cost by subtracting the total cost of producing all but the last unit(s) from the total cost of producing all units including the last one, and dividing the result by the additional output produced by the last unit. Marginal cost can be measured, therefore, by using the formula:

$$\text{marginal cost} = \frac{\text{change in total cost}}{\text{change in output}}$$

We show the marginal costs of production for our hypothetical example in column 4 of Table 7.3 and an equivalent marginal cost schedule is shown graphically in Figure 7.3b. By including marginal cost data on these graphs we can easily identify the minimum cost position.

Finding Minimum Costs

As Figure 7.3b demonstrates, the marginal cost curve first falls and then rises much like the average variable cost and average total cost curves. This should not be surprising: when marginal cost is below average total cost, average total cost falls; when marginal cost is above average total cost, average total cost rises. At the point at which average total costs are neither falling nor rising, marginal cost must then be equal to average total cost. When we represent this graphically, the marginal cost curve intersects the average total cost curve at its minimum.

The same analysis applies to the intersection of the marginal cost curve and the average variable cost curve.

Table 7.3 Marginal and average costs

In the first column, we give the number of workers used per week on a project. In the second column, we give their total product; that is, the output that each specified number of workers can produce in terms of square metres. The third column gives the marginal product. The marginal product is the difference between the output possible with a given number of workers minus the output made possible with one less worker. In the fourth and fifth columns we calculate the marginal costs and average costs per square metre – assuming that each worker is paid £500 a week. For example, the marginal product of a fourth worker is 20 square metres, because with four workers, 140 square metres are produced, but with three workers only 120 are produced. The difference is 20. The marginal cost, therefore, is £25 (£500 ÷ 20), and the average variable cost at the same point is £14.29 (£2,000 ÷ 140).

Input of labour	Total product (output in sq. m per week)	Marginal physical product (in sq. m per week)	Marginal cost (£/sq. m)	Average cost (£/sq. m)
0	0			0
1	20	20	25.00	25.00
2	60	40	12.50	16.67
3	120	60	8.33	12.50
4	140	20	25.00	14.29
5	150	10	50.00	16.67
6	160	10	50.00	18.75
7	165	5	100.00	21.21
8	163	−2	−250.00	24.54

THE CONTRACTOR'S PROJECT COSTS

In most economic textbooks it is explained that producers can progress to determine their selling price from a position of cost analysis. This makes complete sense for most industries where the product already exists, such as motor manufacturing or retailing. In the case of construction, however, the client frequently initiates projects and the contractor has to determine the price before the work is complete. In many instances, there is only one party interested in buying the specific project – this type of market is known as a **monopsony**. This gives construction clients an unusually powerful bargaining position.

What tends to happen is the client invites a selected number of contractors to compete for a project. Contractors responding to this invitation prepare and submit a tender price. Clearly, as part of this process, each contractor will make various estimates before arriving at their final price – but they will never be completely certain of their costs or the time it will take to compete the project satisfactorily.

The Contractor's Bid

It is acknowledged that decisions about whether to bid or not to bid, and at what price, are complex. For example, Shash (1993) has identified more than 50 factors that contractors take into consideration before submitting a bid. Here are Shash's top three – in rank order.

1 **Need for work** – The current state of the contractor's market, such as the firm's position on its cost curves, will be a significant factor in determining the nature of the bid. For example, if fixed costs are not spread over a sufficient volume of work, the firm will be willing to take on work at lower prices than when it is working at capacity. In Shash's words (1993: 111): 'A contractor must secure a designated business volume……to cover his/her operating costs and to realise a reasonable profit.' In other words, the more desperate a firm is for work, the lower its bid price.

2 **Number of competitors** – Firms operating in a crowded, competitive situation have to accept a fair price dictated by others in the market. In contrast, if only a few firms dominate the market, a firm will be able to 'administer' its own price. This comparison between firms 'taking' a price or 'administering' a price will become clearer after the next chapter.

3 **Experience in similar projects** – The degree of complexity in the work required, compared to the firm's experience, is obviously a major determinant. (Shash's research related to the top 300 contractors, and had it broadened out to the lower end of the industry, location of the project would also have been a significant factor in deciding whether to bid.)

LONG-RUN COSTS

The long run is defined as the time during which *full* adjustment can be made to any change in the economic environment – in the long run, all factors of production are variable. For example, in the long run a firm can alter its plant size. There may be many short-run curves as a firm develops over the years, but only one long run. Long-run curves are sometimes called planning curves, and the long run may be regarded as the **planning horizon**.

We start our analysis of long-run cost curves by considering a single firm contemplating the construction of a single plant. The firm has, let us say, three alternative plant sizes from which to choose on the planning horizon. Each particular plant size generates its own short-run average total cost curve. Now that we are talking about the difference between long-run and short-run cost curves, we will label all short-run curves with an S; short-run average (total) cost curves will be labelled SAC, and all long-run average (total) cost curves will be labelled LAC.

Look at Figure 7.4a (see page 114). Here we show three short-run average cost curves for the three (successively larger) plant sizes. Which is the optimal plant size to build? That depends on the anticipated rate of output per unit of time. Assume for a moment that the anticipated rate is Q_1. If plant size 1 is built, the average costs will be C_1. If plant size 2 is built, we see on SAC_2 that the average costs will be C_2, which is greater than C_1. So if the anticipated rate of output is Q_1, the appropriate plant size is the one from which SAC_1 is derived.

Note, however, what happens if the anticipated permanent rate of output per unit of time goes from Q_1 to Q_2. If plant size 1 has been decided upon, average costs will be C_4. However, if plant size 2 had been decided upon, average costs will be C_3, which is clearly less than C_4.

Figure 7.4a Preferable plant size

If the anticipated permanent rate of output per unit time period is Q_1, the optimal plant to build would be the one corresponding to SAC_1 because average costs are lower. However, if the rate of output increases to Q_2, it will be more profitable to have a plant size corresponding to SAC_2, as unit costs can fall to C_3.

Long-run Average Cost Curve

If we make the further assumption that during the development of a firm the entrepreneur is faced with an infinite number of choices of plant size, then we can envisage an infinite number of SAC curves similar to the three in Figure 7.4a. We are not able, of course, to draw an infinite number; but we have drawn quite a few in Figure 7.4b.

By drawing the envelope of these various SAC curves we find the **long-run average cost curve**. To be academically precise, the long run average cost (LAC) curve should be wavy or scalloped, since it follows the path of the SAC curves enclosed. By tradition, however, it is portrayed as being tangent to the minimum point of the SAC curves from which it is derived. Either way, the long-run average cost curve represents the cheapest way to produce various levels of output – provided the entrepreneur is prepared to change the size and design of the firm's plant. Long-run average cost curves are sometimes referred to as **planning curves**.

Figure 7.4b Deriving the long-run average cost curve

If we draw all the possible short-run average curves that correspond to different plant sizes and then draw the envelope to these various curves, $SAC_1 ... SAC_8$, we obtain the long-run average cost curve.

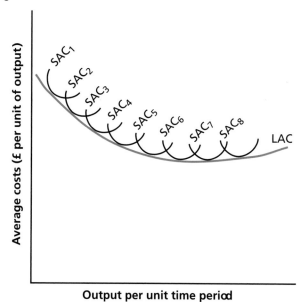

Output per unit time period

Why the Long-run Average Cost Curve is U-shaped

Notice that the long-run average cost curve LAC in Figure 7.4b is U-shaped. It is similar to the U-shape of the short-run average cost curve developed previously in this chapter. However, the reason for the U-shape of the long-run average cost curve is not the same as that for the short-run U-shaped average cost curve. The short-run average cost curve is U-shaped because of the law of diminishing marginal returns. However, that law cannot apply to the long run – in the long run all factors of production are variable, so there is no point of diminishing marginal returns since there is no fixed factor of production. Why, then, does the long-run average cost curve have a U-shape? The reasoning has to do with changes in the scale of operations. When the long-run average cost curve slopes downwards, it means that average costs decrease as output increases. Whenever this happens, the firm is experiencing **economies of scale**. If, on the other hand, the long-run average cost curve is sloping upwards, the firm is incurring increases in average costs as output increases. The firm is said to be experiencing **diseconomies of scale**. There is a third possibility: if long-run average costs do not change with changes in output, the firm is experiencing **constant returns to scale**. In Figure 7.5 we show these three stages. In the first stage the firm is experiencing economies of scale; in the second stage, constant returns to scale; and in the third stage, diseconomies of scale.

Figure 7.5 Economies and diseconomies of scale

Long-run average cost curves will fall when there are economies of scale, as shown in stage one up until Q_1. There will be constant returns to scale when the firm is experiencing output Q_1 to Q_2, as shown in stage two. And, finally, long-run average costs will rise when the firm is experiencing diseconomies of scale, beyond Q_2 in stage three.

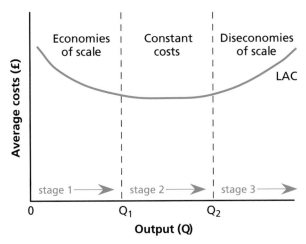

Returns to Scale – In Three Stages

Savings (economies of scale) are possible as firms progress to larger production – that is, increases in output can result in a decrease in average cost. There are five types of scale economies.

- Technical economies: relating to the firm's ability to take full advantage of the capacity of its machinery.
- Managerial economies: as firms grow, they can afford to employ – and benefit from – specialised managers.
- Commercial economies: such as buying in bulk and advertising.
- Financial economies: larger firms have a greater variety of sources for funds and often at favourable rates.
- Risk bearing economies: larger firms may achieve distinct advantages by diversifying into several markets and researching new ones.

When economies of scale are exhausted, constant returns to scale begin. Some economists regard the commencement of this stage as the **minimum efficient scale** (MES) since it represents the lowest rate of output at which long-run average costs are minimised – and no further economies of scale can be achieved in the present time period. The MES is represented by point Q_1 in Figure 7.5.

Clearly, economies of scale are more easily associated with standardised manufactured products. Indeed, in many manufacturing industries a firm has to be big to survive. This is certainly not the case in construction. The unique nature of many construction projects, plus the relatively small size of many construction firms, prevents the industry from realising the full potential of economies of scale.

Cooke (1996: 134) shows very clearly the different capabilities of manufacturing and construction firms to benefit from economies of scale by using a diagram similar to Figure 7.6. In Cooke's analysis the long-run average cost curve LAC_1 represents the average costs of production of a typical manufacturing firm able to exploit economies of scale. In this case, the MES is at point Q_2. In contrast, LAC_2 – positioned to the left of and above LAC_1 – denotes the long-run average costs associated with projects with lower levels of standardisation, such as those experienced by the typical firm in construction. On LAC_2, a firm can only achieve economies of scale to output Q_1 – so clearly it would be beneficial to reorganise production to be on LAC_1. This is precisely what the contractor O'Rourke achieved by its acquisition of Laing Construction in 2001. Within a year O'Rourke was in a position to offer a range of standardised, prefabricated components that could be configured to create buildings to suit individual client's requirements. This development was only possible due to O'Rourke's increased scale of operation. In effect it had moved from operating along LAC_2 to the more efficient LAC_1 schedule. These outcomes are often the rationale that lie behind mergers and acquisitions – but these achievements are still the exceptions rather than the rule in the construction industry.

Figure 7.6 Economies of scale in the construction sector

LAC_1 represents the average cost of production for a firm able to exploit economies of scale, such as a manufacturer. In comparison, LAC_2 is the average cost of production for a contractor unable to take full advantage of economies of scale, due to the unique nature of each unit of output.

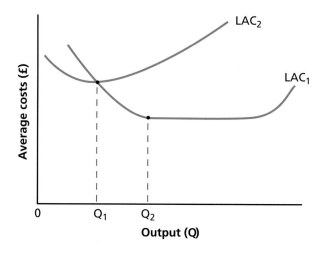

To sum up, the general rule seems to be that economies of scale, of any kind, have a limited role in construction. In fact, Patricia Hillebrandt consistently emphasised in each edition of her textbook that 'many construction firms are actually on a long-run increasing cost curve' (Hillebrandt, 1974: 121; 1985: 116 and 2000: 121). She attributes the problem to the localised nature of small traditional construction firms. For example, a firm needing to dig ditches might hire workers on a job basis. For jobs requiring up to ten workers, it simply hires the workers and gives them each a shovel to dig ditches. However, for jobs requiring ten or more workers, the firm may feel it is also necessary to hire an overseer to co-ordinate the ditch-digging effort. Thus, perhaps, constant returns to scale remain until ten workers are employed, then decreasing returns to scale set in. As the layers of supervision grow, the costs of communication grow more than proportionately; hence, the average cost per unit of output will start to increase. In the terms used in Figure 7.5, firms can run quite quickly into diseconomies of scale as they grow in size if they are not careful.

According to Hillebrandt's analysis these diseconomies occur very soon in construction for two main reasons. First, most firms can only substantially increase their turnover by extending their catchment area, which in turn increases costs of transport and supervision. Second, there is the indivisible nature of entrepreneurial ability that causes the decision-making process to clog up as firms increase their scale of operations. This second argument is well rehearsed in the traditional business literature: many analysts have observed a crisis point in a firm's growth when it becomes too big for the directors of the firms to continue to exercise effective personal control yet too small to afford to recruit extra expertise. The precise point that this occurs depends on the management calibre of the existing directors.

A more current debate, captured in the research of Pan et al. (2008) and Chiang et al (2008) focuses on the new opportunities offered by **off-site production (OSP)** where buildings, structures or parts are manufactured or pre-assembled prior to their installation on site. This offers contractors the opportunities to take advantage of a manufacturing environment, with opportunities to improve on quality, time, cost, productivity and safety. Yet surveys of the industry in the UK and Hong Kong suggest that the difficulties in achieving economies of scale continue to be a significant constraint, restricting construction to traditional on-site approaches and cost schedules (Pan et al. 2008). The debate regarding whether traditional organisational problems or modern technological barriers limits a construction firm are examined in Reading 2 (see pages 139–141) where the option of off-site production to reduce average costs per unit is reviewed further.

The Minimum Efficient Scale and Levels of Competition

The manufacturing sector seems to be a constant reference point in these debates where, relatively speaking, state-of-the-art savings and economies of scale are more easily achieved. In fact, the conceptual idea of a minimum efficient scale (MES) first originated from a comprehensive study into the shape of long-run average cost curves in British manufacturing industry. This was defined as 'the minimum scale beyond which any possible subsequent doubling in scale would reduce total average unit cost by less than five per cent' (Pratten 1971: 278).

One of the uses of the minimum efficient scale (MES) is that it gives a rough measure of the degree of competition in an industry. If the level of output associated with the minimum efficient scale is small relative to the total industry demand, the degree of competition in that industry is likely to be high since there is room for many efficient firms. Conversely, when the level of output associated with the MES is large relative to industry demand, the degree of competition is likely to be small. For example, the markets in oil and gas extraction, air transport, automobile manufacturing and tobacco are dominated by a few very large businesses. These industries have a high **concentration ratio**, and many have the characteristics of oligopolies. In stark comparison, some industries are typified by many small businesses, some of which are run by the self-employed. These industries are said to have low concentration ratios. Construction is a prime example of this type of industry and these markets are characterised as contestable. (Types of market competition will be discussed further in Chapter 8.) Economists maintain that the number of firms comprising an industry is determined by the MES, as they claim that an industry is 'efficient' when the amount of resources used to produce the total output is at a minimum.

One way to estimate the significance of economies of scale to specific industries, therefore, is to determine the gross value added (the turnover), the percentage of total output, the employment or the profit accounted for by, say, the top four or five firms in an industry. These are the common measures used to estimate the concentration ratio for an industry. In the statistical material published by the Office for National Statistics (ONS) gross value added and output are the key variables calculated for the top five businesses; and these are used to produce a five-firm concentration ratio. In other words, the volume of output and the gross value added (GVA) of the top five businesses are measured as a proportion of the total output or total turnover of the industry. The series is based on the *Annual Business Enquiry and Input-Output Analyses*. Note, however, before considering examples, that the precise definition of any specific industry is somewhat arbitrary and, consequently, as we narrow the definition of an industry the concentration ratio rises, and vice versa. Also conventional measures include only domestic production, and if foreign competition is a significant presence, the actual market concentration will be much lower than indicated by the ratios.

Table 7.4 (on page 120) gives the five-firm concentration ratio for four industries. As noted, the more precise the definition of an industry the higher the ratio. Consequently, the broad industrial classification of construction produces a very low concentration ratio, as both total production and turnover (gross value added) are dominated by small businesses. In fact, as we pointed out in Chapter 6 there are more than 186,000 enterprises comprising the construction industry – responsible for all the gross value added and output of the UK construction firms. However, if we define house building as a separate industrial sector then it becomes

Table 7.4 Concentration ratios by industry (in 2004)

Industry	Five-firm concentration ratio	
	GVA (%)	Output (%)
Construction	5	5
Iron and steel	48	61
Cement, lime and plaster	74	71
Metal boiler and radiator	51	45

Source: Input-output analysis (ONS, 2006: Table 8.31)

a more concentrated market – the top five house builders account for more than 65 per cent of the gross value added by all the house builders across Great Britain. This increases the opportunities to benefit from economies of scale and changes the nature of competition. Indeed, house building at the regional and local level often allows businesses to act like firms operating in an oligopoly, as the number of competing firms at this level will be quite small. The bigger house building firms in an area should be able to control the price and quality expected in that marketplace.

Table 7.4 also shows that the five-firm concentration ratio for iron and steel output is approximately 61 per cent. This seemingly high ratio is to be expected from a specific production process that requires heavy plant and machinery. It has, however, fallen significantly. In 1997, the five-firm concentration ratio in this industry was 95 per cent. This change can be accounted for by the structural changes that iron and steel have experienced during the last decade. As noted above, the balance of domestic production in relation to foreign imports and new technologies will also distort these calculations. Finally, it is interesting to observe that the ratio between the output and the gross value added by the top five firms in the iron and steel sector is significantly different to all the other industrial sectors shown in Table 7.4. This suggests that the higher value, high-quality iron and steel is produced by smaller specialist firms in this industrial sector.

Key Points 7.4

❑ The long-run average cost curve is derived by drawing a line tangent to a series of short-run average cost curves, each corresponding to a different plant size.

❑ A firm can experience economies of scale, diseconomies of scale and constant returns to scale, depending on whether the long-run average cost curve slopes downwards, upwards or is horizontal (flat).

❑ Economies of scale occur when all factors of production are increased and average costs fall. There are five reasons for economies of scale, relating to (a) managerial economies, (b) commercial economies, (c) financial economies, (d) technical economies and (e) risk-bearing economies.

❑ The minimum efficient scale occurs at the lowest rate of output at which long-run average costs are minimised.

❑ Construction firms may experience diseconomies of scale because of limits to the efficient functioning of management, the fragmented nature of the industry and the difficulties experienced in achieving economies of scale.

❑ Concentration ratios estimate the importance of the largest firms in an industrial sector and give some indication of the level of competition and opportunities to benefit from economies of scale.

EXTERNAL ECONOMIES OF SCALE

The economies of scale we have discussed so far in this chapter are internal to the firm, and could be described as the direct result of individual company policy. That is to say, they do not depend on what other firms are doing or what is happening in the economy. They are formally referred to as **internal economies** (or diseconomies) of scale. This is to distinguish them from **external economies of scale**, which benefit all firms in an industry – regardless of their individual size or policy. In the case of construction, external economies are more important than internal economies of scale. In most countries, the construction industry represents one of the largest sectors – yet is usually dominated by many fragmented firms. The external economies in construction are, therefore, in effect the by-product of being a firm involved in a large industrial sector.

When any industry expands, *all* firms in the industry benefit in ways that normally lead to savings for *all* the firms involved. Firms get the opportunity to buy in services more easily; firms can combine to fund research and/or training; firms often become more specialised; trade associations may form; professional bodies can emerge to represent their members; and specialised journals may be started to report on best practice. In construction, there are examples of all these developments.

As we mentioned in Chapter 1, during the last 40 years there have been several government reports seeking to improve the efficiency of the industry as a whole. *Rethinking Construction*, the report by Sir John Egan (1998), set a national agenda to improve the performance and culture of the construction industry. Targets relevant to our discussion in this chapter included a 10 per cent reduction in annual costs and making tendering more efficient. To achieve these targets, the report suggested that the industry adopts the use of Internet services and collaborates more effectively. Obviously using an Internet package to process transactions significantly reduces administration costs and resolves any unnecessary duplication of effort. The development of two dedicated portals in 2001 – BuildOnline and Asite – provided the mechanism to deliver some of these solutions. These tools enable the procurement of materials through online catalogues, the standardisation of tendering documents and the sharing of information. All firms in the industry now have an improved knowledge of costs, time and quality. Ultimately, it is hoped that information technology will put an end to the wasteful process of competitive tendering.

Another good example of external economies of scale is the various initiatives undertaken by the government on behalf of an industry. For example, Constructing Excellence is a government sponsored organisation that offers firms in the industry opportunities to learn how to improve practice through various case studies and demonstration projects. Constructionline is a free database, launched in March 2006, that gathers information on contractors and consultants to provide public sector procurers (such as central government departments, local authorities, universities, further education colleges and NHS trusts) with quick access to a list of fully accredited suppliers. The list covers the full spectrum of construction activities from architecture to demolition and includes businesses ranging in size from small specialists to the largest main contractors. A final example of an external economy provided by the government is UK Trade and Investment, an organisation that is dedicated to helping UK companies grow their business internationally. This is done through a global network of staff working in the UK and abroad to help any company wanting to export. A relevant example of its work was the publication of a report that provided full details (including barriers and incentives) of every pending major transport project in India (UK Trade and Investment 2007). The report highlighted the incredible potential for contractors seeking to expand into the Indian market, as it pointed out that the Indian government had earmarked £4.5 billion to upgrade 25 airports, £27 billion for roads and highways development, £6.3 billion for 276 port projects and a further £10 billion to upgrade its old and large railway system. Such a report could clearly benefit major UK construction companies, especially as the Indian government is interested in those businesses with experience in public private partnerships, as it regards these as a way of quickly building up the nation's infrastructure. To sum up, external economies, regardless of whether they are provided by trade associations, professional organisations or the government, help to improve overall industry performance.

These developments represent good examples of external economies of scale, as in each case, industry-wide initiatives encourage the reduction of each participating firm's costs. In terms of Figure 7.5, the industry as a whole should experience a downward shift in its long-run average cost curve reflecting a decrease in costs at every level of output.

Of course, external change will not always benefit the member firms of a particular industry. For example, as the construction industry grows larger, shortages of specific materials, land and /or skilled labour may occur. This would push up costs per unit of output across the whole industry – or, in the terms employed above, the industry would experience external diseconomies of scale.

A Final Note on Techniques

Before progressing further with the theory of the construction firm, it may be useful to emphasise three points which should help you avoid any confusion when studying other economic textbooks.

1 By now you should understand how economists use cost curve diagrams to provide a summary of their ideas. You should realise, therefore, that these diagrams are never intended to be more than a visual image – an aide-memoire. They never form a precise reflection of a particular firm.

2 Most microeconomic theory has been developed with manufacturing as the focus and it does not, therefore, always smoothly translate to the construction industry. As Ive and Gruneberg (2000: 150) emphasise, economic theory relates to the manufacturing of mass-produced products, whereas construction is oriented towards individual projects that tend to be 'one-off productions' (with the possible exception of housing). Indeed, in most mainstream textbooks on economic theory the construction industry is not even indexed.

3 Remember that we commence from the basic assumption that all firms in all sectors seek to maximise their profits. This is an important cornerstone of economic theory. Economists pursue this objective with an academic zeal that regards even the marginal detail as crucially significant. This will be developed further in the next chapter.

Key Points 7.5

❑ External economies of scale relate to a whole industry; whereas internal economies of scale arise from the growth of one firm, regardless of what is happening to other firms.

❑ Good examples of external economies of scale arise from the work of government departments that seek to deliver a marked improvement in the business performance of construction firms at home and abroad.

❑ The cost-curve techniques associated with the theory of the firm have been developed for, and particularly apply to, the profit maximising manufacturer.

8 Types of Market Structure in the Construction Industry

In order to understand the precise relationship between output, revenue and price, a firm has to know the structure of the market or industry into which it is selling its product. There are various market structures. At one extreme, there is a **monopoly** where one producer dominates the market and controls the price and output decisions. At the other extreme, both buyers and sellers correctly assume that they cannot affect market price – this market structure is known as **perfect competition**. Most firms involved in construction are engaged in market structures between these two theoretical extremes. In the language of textbook economics, they are involved in **imperfectly competitive markets** – where the clients and contractors have to take into account how their individual actions will affect the market price. (Here, it might be useful to review Key Points 6.1.) We shall examine these real-life scenarios in more detail once we have set up a reference point for the discussion.

THE PURPOSE OF PERFECT COMPETITION

To begin, we consider in some detail the hypothetical scenario of a perfectly competitive market. Although no real industry actually operates in such a market, it provides an important reference point for economists. In the case of construction, there are also some interesting parallels in reality. The perfectly competitive market acts as a benchmark from which other market situations can be judged. As we will show, an optimum allocation of resources arises from perfect competition because every firm is producing at minimum unit cost. Consequently, we will be able to develop an understanding of what is meant by a 'fair' price, a 'normal profit' and an 'efficient' industry.

The Characteristics of Perfect Competition

The term perfect competition relates to a specific model market structure with these defining characteristics.

- The product sold by firms in the industry is homogeneous. This means that the product sold by each firm in the industry is a perfect substitute for the product sold by every other firm. In other words, buyers are able to choose a product from a large number of sellers in the knowledge that it is essentially the same. The product is thus not in any sense differentiated regardless of the source of supply.
- Any firm can enter or exit the industry without serious impediments. Resources must also be able to move in and out of the industry unimpeded; without, for example, government legislation preventing any resource mobility.
- There must be a large number of buyers and sellers. When this is the case, no single buyer or seller has any significant influence on price. Large numbers of buyers and sellers also means that they will be acting independently.

• There must be complete information available to both buyers and sellers about market prices, product quality and cost conditions.

Now that we have defined the characteristics of a perfectly competitive market structure, we consider the position of an individual firm. We define a **perfectly competitive firm** as:

> one that is such a small part of the total industry in which it operates that it cannot significantly affect the price of the product in question.

This means that each firm in the industry is a **price-taker** – it takes the price as something that is beyond its individual control.

How does a situation arise in which firms regard prices as set by forces outside their control? The answer is that even though every firm, by definition, sets its own prices, it must always considers the prices of its competitors. The firm in a perfectly competitive situation finds that it will eventually have no customers if it sets its price above the competitive price. Let us now see what the demand curve of an individual firm in a competitive industry looks like.

Single-Firm Demand Curve

We have already discussed the characteristics of demand schedules (for example, see Key Points 4.1). Figure 8.1 presents the hypothetical market demand schedule faced by any producer or contractor, who, we assume, controls only a very small part of the total market. This is how we characterise the demand schedule for a purely

Figure 8.1 The demand curve for an individual firm in a perfectly competitive market

We assume that the individual producer represents such a small part of the total market that it cannot influence the price. The firm accepts the price as given. At the going market price it faces a horizontal demand curve. If it raises its price, even by one penny, it will sell nothing. Conversely, the firm would be foolish to lower its price because it can sell all that it can produce at the market price. The firm is a price-taker and its demand curve is described as being perfectly elastic.

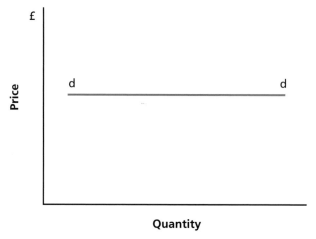

competitive firm – it is a horizontal line at the going market price. It is a completely elastic demand curve – by raising its price by one penny, the individual firm would lose all its business.

At the market price, demand is **perfectly elastic**; the firm can sell as much output as it wants, providing it does not alter the price. If the firm were to raise its price, consumers will buy the same skill, product or service from another producer.

Key Points 8.1

❑ There are various market structures. Examples include a monopoly in which one producer alone controls price and output decisions, and perfect competition in which no single producer can control the market.

❑ The perfectly competitive market is a hypothetical extreme that acts as a benchmark by which real (imperfect) markets can be judged.

❑ The hypothetical model of perfect competition has four main characteristics: (a) homogeneous product, (b) freedom of entry and exit, (c) large number of buyers and sellers, and (d) full information.

❑ A perfectly competitive firm is a price-taker. It takes price as given. It can sell all that it wants at the going market price. The demand curve facing a perfect competitor is a horizontal line at the market price.

HOW MUCH DOES THE PERFECT COMPETITOR PRODUCE?

We have established that a perfect competitor has to accept the given price of the product. If the firm raises its price, it sells nothing. If it lowers its price, it makes less money per unit sold than it otherwise could. The firm has only one decision variable left: how much should it produce? We will apply our model of the firm to answer this question. We shall use the **profit-maximisation model** and assume that firms, whether competitive or monopolistic, will attempt to maximise their total profits – that is, they will seek to maximise the positive difference between total revenue and total costs. (It may also help here to review the distinction between accounting and economic profits, see Key Points 7.1.)

Total Revenue

Every firm has to consider its total revenue. Total revenue is defined as the quantity sold multiplied by the price; or, expressed using the notation employed in some texts, TR = P x Q. This is also the same as total receipts from the sale of output.

In Figure 8.2 (see page 128) we assume that the firm is one of many comprising the total market, so it can sell all it produces at a given price. Thus, the total revenue curve is presented as a straight line. For every unit of sales, total revenue is increased by proportionally the same amount.

Total Costs

Revenue is only one side of the picture. **Total costs** must also be considered. Notice that when we plot total costs in Figure 8.2 the curve is not a straight line, but a wavy line, due to the existence of increasing and decreasing returns which we alluded to in Chapter 7. When the total cost curve is above the total revenue curve, the firm is experiencing losses. When it is below the total revenue curve, the firm is making profits. Where the two curves intersect represents breakeven points. Note that by reducing total costs, firms can make bigger profits.

Figure 8.2 Finding a profit-maximising position

The straight black line represents total revenue, as each unit is sold for the same price. Total costs, represented by the orange line, first exceed total revenues and a loss is made; then they become less than total revenue and a profit is made. We find maximum profits at the point where total revenues exceed total costs by the largest amount.

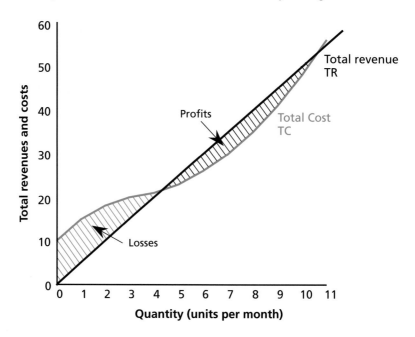

COMPARING TOTAL COSTS WITH TOTAL REVENUE

By comparing total costs with total revenue, it is possible to calculate the number of units that the individual competitive firm should aim to produce per month. Clearly, the firm will maximise profits at that place on the graph where the total revenue curve exceeds the total cost curve by the greatest amount. In Figure 8.2, that occurs at a rate of output and sales of either seven or eight units per month; this rate may be called the profit-maximising rate of production.

Marginal Analysis

Another way to find the profit-maximising rate of production for a firm is by marginal analysis. This method involves making a detailed study of marginal revenue and marginal costs. The concept of **marginal cost** has already been introduced in Chapter 7. It was defined as the change in total cost due to a one-unit change in production. The resulting schedule of costs was based on the law of diminishing returns: at first costs fall and then they begin to rise. Some example calculations for a marginal cost schedule were presented in column 4 of Table 7.3. This leaves **marginal revenue** to be clarified.

Marginal Revenue

Marginal revenue represents the increment in total revenue attributable to selling one additional unit of product. For example, if selling an extra unit of construction activity increases a contractor's total revenue from £1,800 to £2,100, the marginal revenue equals £300. Hence, marginal revenue may be calculated by using the formula:

$$\text{marginal revenue} = \frac{\text{change in total revenue}}{\text{change in output}}$$

In any market structure, therefore, marginal revenue is closely related to price. In fact in a perfectly competitive market, the marginal revenue curve is exactly equivalent to the price line or, in other words, to the individual firm's demand curve, since the firm can sell all of its output (including the last unit of output) at the market price.

COMPARING MARGINAL COST WITH MARGINAL REVENUE

Obviously, if the marginal revenue from a unit increase in output is greater than the marginal cost, it would seem rational for the profit-maximising firm to produce that unit of output. Conversely, if the marginal cost of an extra unit of output exceeds its marginal revenue, it would be produced at a loss and, therefore, it would be inappropriate for the profit-maximising producer to produce that unit of output. In fact, all firms have a clear incentive to produce and sell right up to the point at which the revenue received from selling one more unit of output equals the additional cost incurred in producing that unit. If the firm chooses to stop output before this point, it will not have maximised profits: it will not have squeezed the pips until they squeak. The profit maximiser should not be satisfied until the last penny of profit has been earned. This will only be achieved at the point where marginal costs equal marginal revenue. This decision rule is represented by point E in Figure 8.3 (see page 130).

Figure 8.3 Long-run perfectly competitive equilibrium

In the long run, perfectly competitive firms move towards a position at which marginal revenue equals marginal cost and average total costs. In short, 'where everything is equal' – represented by point E.

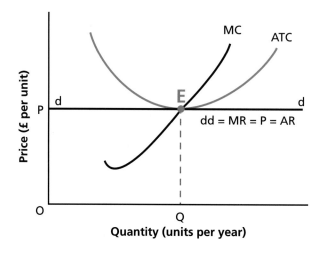

Key Points 8.2

❑ Profit is maximised at the rate of output where the positive difference between total revenue and total costs is greatest.

❑ Using marginal analysis, the profit-maximising firm will produce at a rate of output at which marginal revenue equals marginal cost.

❑ Profit-maximising rules apply to all types of market structure.

TOWARDS THE NOTION OF AN EFFICIENT INDUSTRY

To consider an entire industry, the cost and revenue schedules of all its constituent firms need to be aggregated. For an industry with a perfectly competitive market structure, this is not a problem since the costs and revenue for each firm are identical. This theoretical extreme will prove relevant when we consider the notions of efficiency that exist in reality. It should also prove to be useful to those concerned with the performance of firms in the construction sector. Most firms in construction do seek to maximise their profits, they often have a large degree of freedom to enter and exit the various activities, and they rarely set their prices without regard to the terms expressed by their competitors.

Short-run Versus Long-run Profits

Economic theory suggests that in the long run all the firms in a competitive industry earn normal profits (as defined and discussed in Chapter 7). In the short run, however, which may represent a considerable length of time in construction markets, some firms will sell their product well above their minimum average costs and make **supernormal profits**. As a result, more firms will be enticed into the sector to get a slice of the action. In time, this increased competition will force the equilibrium price of the product down, until each firm is making only normal profit. This situation is shown in Figure 8.3.

In the long run, the perfectly competitive firm/industry finds itself producing at a rate Q. At that rate of output, the price is just equal to the minimum average total cost. Obviously, it is possible for supernormal profits to cause too many firms to enter the market, in which case the market price would fall below P and firms would make a loss – **subnormal profits**. Firms would then leave the industry and the decrease in supply would cause market prices to rise again to P. In this sense, perfect competition results in no 'waste' in the production system. Goods and services are produced using the least costly combination of resources. This is an important attribute of a perfectly competitive long-run equilibrium, particularly when we wish to compare the market structures that are less than perfectly competitive.

Key Points 8.3

❏ In the short run, the perfectly competitive firm can make supernormal profits.

❏ In a perfectly competitive market, new firms entering the industry will absorb supernormal profits.

❏ In the long run, a perfectly competitive firm (industry) produces at the point where P = MR = MC = ATC.

MARKET STRUCTURES THAT TYPIFY THE CONSTRUCTION INDUSTRY

So far we have discussed in some detail the hypothetical market structure of perfect competition, in which there are numerous firms that produce the same product and have no influence over price: they are *price-takers*. In this section, we analyse the actual markets in which construction firms are engaged. These are also generally competitive, in the sense that there are usually many firms producing an insignificant part of total construction output, and there are certainly no formal restrictions preventing firms from participating. Most firms comprising the construction industry do, however, have some control over price. In fact, we shall conclude that, in general terms, markets representing construction are often representative of either monopolistic competition or oligopoly and, in practice therefore, there is an element of control over *price-making*.

In making this comparison between the world of perfect theory and actual practice, it is important to remember that the construction industry, as a whole, consists of many different markets – some are defined by a specific service or product; others by the size and complexity of contracts awarded in the market, or by the geographical location of the market. Consequently, we should not aim to pigeonhole all firms into one model of market behaviour. However, the following analysis represents a quick tour through some generalised principles that apply to significant parts of the construction industry. It may help your comprehension if you have some specific construction firms in mind as you proceed.

Monopolistic Competition

In reality, most markets are far from perfect. For example, in any construction market contractors, subcontractors and material producers will try to obtain some monopoly advantages by distinguishing their firm's product from that of their competitors. They may do this by somehow implying – or, indeed, achieving – better quality and/or reliability. In these types of market, firms can earn above normal profits – but only for a short while, because other firms in the market will respond by producing similar products. This keeps the market very competitive, and constrains long-term profits.

This model of behaviour is known as **monopolistic competition** as each firm can easily achieve a degree of local monopoly but is ultimately restricted by the presence of many competing firms. As we have suggested, this type of competition is well exemplified by firms in the construction industry: firms tend to be highly fragmented across the country but are somehow constrained by the potential competition of similar firms in neighbouring towns.

Oligopoly

In the strictest sense of the word, **oligopoly** is where a few sellers compete for the entire market. Blue Circle, for example, accounts for approximately half of the supply of cement to building firms in the UK and two other firms, Rugby and Castle, make most of the remaining cement. In this kind of market each firm has enough power to avoid being a price-taker – but they are still subject to a sufficient amount of competition to know the market is not entirely under their control. In other words, firms in oligopolistic markets will price their produce or service according to how they think competitors will react. This leaves them faced with the dilemma of not knowing whether to compete or co-operate. In short, the world of oligopoly is one of uncertainty.

To paraphrase Lipsey and Crystal (1995: 264), firms in an oligopolistic industry will make more profits if they agree to co-operate as a group; however, if one firm deviates from the agreement and/or becomes aggressively competitive, it stands to make more profit for itself.

When firms agree to co-operate to raise profits it is called **collusion**. According to anecdotal evidence from several groups of postgraduate students working in the industry, collusion is common practice in contractual agreements across the whole breadth of construction. This recent experience was confirmed by the extensive investigations carried out by the Office of Fair Trading in 2008 that led to more than 100 British construction companies being accused of rigging bids. Hillebrandt's theoretical analysis also emphasised that many firms in the construction industry follow some form of oligopolistic behaviour. This was a dominant message throughout her writing career (1974: 155–7; 2000: 153–5). Since collusive agreements of one type or another tend to typify many transactions in the industry, we need to understand them in some detail.

THE HYPOTHESIS OF QUALIFIED JOINT PROFIT MAXIMISATION

As long ago as 1776, Adam Smith, the founder of modern economics, alluded to collusion. In commenting upon contracts between rival firms, he showed deep suspicion. He observed that 'people of the same trade seldom meet together for fun and merriment, but the conversation ends in a conspiracy against the public, or in some contrivance to raise prices'.

It was not until 170 years later that this statement was elaborated into a hypothesis in the writing of Professor William Fellner. He explained how some kind of agreement between competing firms of the same trade was inevitable. The agreements, however, were not necessarily formal. In fact he distinguished between two types of agreement: 'explicit agreements' and 'quasi-agreements'. Explicit agreements would nowadays be regarded as 'overt' or 'covert' collusion, depending on whether the agreement is open or secret. In contrast, a 'quasi-agreement' is far less formal – it is what Fellner describes as somehow representing a 'spontaneous co-ordination'. An example of a quasi-agreement would be the kind of unwritten rule that says all firms should take their price from the most dominant company in the market to avoid rounds of price-cutting. This latter type of arrangement, where no formal agreement actually occurs, is now more commonly referred to as tacit collusion.

Although collusive behaviour of any form is usually frowned upon, it is entirely understandable that firms may try to act in a common manner to protect or promote common interests. The importance of Fellner's work is that it shows that agreements to protect common interests do not require an explicit arrangement. As Fellner (1949: 16) noted: 'The difference between "true" agreement and quasi-agreement is that the former requires direct contact while the latter does not.' The basic aim of his hypothesis was to explore how rival firms in an oligopolistic type market resolved the conflict of knowing whether to:

- compete with rivals to gain as large a share of the potential profits as possible
- co-operate with rivals to maximise their joint profits.

The relative strength of these two forces – which operate concurrently – varies from industry to industry as the following five market characteristics suggest. As you read through them try to think about how many may apply to a specific market within the construction industry. In formal terms, that is, how many of these five market characteristics could lead to joint profit-maximising behaviour in a specific construction market.

1   *There are very few firms.* They know each other well enough to understand that one of them cannot gain sales without inducing retaliation. So some agreement to co-ordinate their policies may be reached.

2   *The firms produce similar products.* As a result, it is difficult to gain a specific advantage in the market. In such a situation, firms may prefer some form of joint effort in preference to the cut-throat behaviour necessary to take customers away from each other.

3   *There is a dominant firm.* Other firms may look to the dominant one for its judgement about market conditions and take its lead on prices. In short, the dominant firm becomes a reference point and the focus for tacit agreement.

4   *The firms have very similar average costs.* In this case it is unlikely that firms will enter into price competition. Rivalry could break out in other forms, unless some joint agreement is reached to maximise profits.

5   *New entrants face significant barriers to entry.* The theory of perfect competition suggests that high profits in an existing market will attract new entrants and, as a result, prices and profits reduce. This profit-damaging activity is less likely to occur if some agreement between the existing firms has been made to prevent other firms breaking into the market.

COLLUSION – RIGGED PRICES

It is difficult to gauge the extent of qualified joint profit-maximising agreements, especially as in most countries it tends to be illegal. There are, however, countless opportunities that one can envisage and the examples that are talked about probably only represent the tip of the iceberg. As Fellner (1949: 16) stated when setting up his hypothesis: 'In the real world, quasi-agreement may shade over into true agreement by gradations.' In other words, it is a short hop from the activity of informal or tacit collusive oligopolistic behaviour to the more formal rigging of markets. Either practice raises questions about a lack of fairness and transparency – and leads to imperfect competition of one sort or another.

Clearly it becomes more feasible to contemplate some sort of agreement to divide the market and raise the profit level if there are relatively few contractors in a market. But, as the investigation carried out during 2008 by the OFT into anti-competitive agreements showed, this does not mean that collusion is rare in the cnstruction industry. This investigation uncovered 240 infringements across 112 construction companies. Firms were allegedly engaged in bid rigging activities and in **cover pricing.**

Cover pricing describes an age-old practice where one contractor asks another contractor on the same tender list to quote a price which will be above that quoted by the interested contractor. In other words, it involves one or more bidders colluding with a competitor during the tender process to put in prices which are intended to be too high to win the contract. As a result, the tendering authority is left with a false impression of the level of competition and this could easily lead it to pay inflated prices for the work. In short, those involved in a bid rigging cartel work together to decide who will win a contract and at what price. In extreme cases firms may even 'pay off' those who agree not to tender a bid, although the more common arrangement seems to be where the firm providing the 'cover price' does so in exchange for a similar favour when future contracts arrive in the market. In these cases, the group of contractors are virtually, by concerted action, forming themselves into the structure of an oligopoly so that they can raise the level of profits and share them around on a rota basis.

There have certainly been countless situations in the construction industry where some form of agreement is alleged to have been operated. Apart from being illegal, the durability of such agreements is severely limited by the free entry to the industry and its markets. However, Gruneberg (2008) defends the practice of cover pricing on the grounds that construction firms are providing an identical service and tend to operate in a market that offers relatively low profit margins. Consequently he does not find it surprising that companies respond by seeking covert and informal agreements to manage their work load efficiently. He puts forward the argument that companies in other sectors can manage their throughput by adjusting their levels of output, whereas construction firms on tender lists have to meet their clients' requests. So when a construction firm is operating at near full capacity it might tender a cover price simply to remain on a list of recognised contractors for future projects. This makes sense when one remembers that full tendering is a rigorous and costly exercise, and the contractors that don't win the contract must also meet the costs of tendering.

Regardless, more than 25 per cent of all the European Commission decisions on collusion relate to the construction industry (Gunny 2005). The building materials also has examples of this type of practice and these are examined further in Reading 3 (see page 142). The intriguing aspect is that although in statistical terms the construction industry comprises many thousands of small competitive firms, it is fragmented by trades and regions – and in many of these market segments there are examples of a local monopoly or a local oligopoly with a price leader.

THE THEORY OF GAMES

In recent years economists have begun to use a branch of mathematics called **game theory** to study collusion. In this work, the competition between firms is analysed as a game. Each firm decides its own game plan in terms of price and output, but realises that the result or success of its strategy depends upon the action of its opponents (the competing firms).

In 1994 three game theorists shared the Nobel prize for economics and since then examples of the application of game theory have proliferated in the mainstream texts. Two common examples are the prisoners' dilemma and the zero-sum game. The prisoners' dilemma is used to demonstrate that competing firms have conflicts of interest (like the conspirators in a crime who are interviewed separately) that may or may not be resolved by some kind of tacit agreement. The zero-sum game is used to describe a situation in which the total winnings are fixed – some must lose, if others win. This is similar to competing for work through the tendering process introduced in chapter 5. (Unless, of course, there is some agreement to share the winnings on a rota basis!)

Contestable Markets

Most of the market behaviour we have described in this chapter has been concerned with the actions of firms *inside* the market. The theory of **contestable markets** focuses on the possibility of firms entering the market from the *outside* and the effect that this potential competition has on the behaviour of the firms already inside the market.

The relevance of this theory is that markets do not have to contain many firms for profits to be held near the competitive level – the threat of a potential new entrant is sufficient to constrain prices. In very general terms, research based on the five-year period 1990–1994 concluded that construction markets were contestable and that any high profits due to market powers were unlikely to persist (Ball et al. 2000). During the subsequent decade (1994–2004), however, the contestability of construction markets becomes questionable. At the bottom end of the market, registration schemes have made it more difficult for construction firms to enter the industry; at the top end, an increasing trend for partnering arrangements has also reduced the number of firms able to compete. In short, as **barriers to entry** emerge, competition levels in the industry are becoming less apparent.

Key Points 8.4

❑ The construction industry comprises such a wide diversity of firms that it is impossible to categorise them all within one type of market structure.

❑ Monopolistic competition is a market structure that lies between pure monopoly and perfect competition.

❑ Oligopoly refers to a market structure in which there are just a few firms that are highly interdependent. In very generalised terms, construction firms can be seen to follow this type of market behaviour.

❑ Because the world of oligopoly is one of uncertainty, there are incentives to try to collude, and there are several types of agreements that can exist between firms, including tacit, covert or overt arrangements.

❑ The actions of rival firms may affect the size of the profits of all firms in a market. This is highlighted by the hypothesis of qualified joint profit maximisation and game theory.

❑ A market is perfectly contestable if there are no barriers to entry or exit. The absence of barriers means any new entrant can compete with existing firms in the market.

RESOURCE ALLOCATION AND SUSTAINABILITY

This chapter has demonstrated the theoretical significance of free markets. It has shown how resources are used more efficiently if there is freedom of information, and firms know about each other's procedures. The analysis has clearly emphasised that as markets become more competitive there is a greater likelihood that society's welfare will be maximised. To paraphrase Adam Smith's writing of two centuries ago: as people follow the signals of the invisible hand of the competitive market in pursuit of their own interest they unintentionally also promote public interest. Samuelson and Nordhaus (2005: 283) express the same sentiment in a more modern style: 'The rough and tumble of market competition is a potent force for raising output and living standards.'

Table 8.1 Suspicious bidding patterns

✓	Bids received at the same time containing similar or unusual wording
✓	Bidder betraying discussions with others or with knowledge of previous bids
✓	Bids containing less detail than expected
✓	Likely bidder failing to submit a bid
✓	Lowest bidder not taking the contract
✓	Bids that drop on the entry of a new or infrequent bidder
✓	Successful bidder later subcontracting work to a supplier that bid higher
✓	Suspiciously high bids without logical cost difference (e.g. delivery distances)

Source: Adapted from OGC (2006: 18)

Unfortunately, however, there is a dilemma, as left to their own devices firms attempt to rig markets in their favour. Collusive agreements and oligopolistic market structures lead to higher profit levels. There is a tension as firms tend to favour the higher profits associated with imperfect markets and governments are keen to foster the resource efficiency associated with competitive markets. Resource efficiency is particularly relevant for governments pursuing a sustainable development agenda. Governments, therefore, prefer that firms do not make arrangements to restrict competition and control prices.

Indeed the Office of Government Commerce (OGC), the National Audit Office and the Office of Fair Trading each has a responsibility to assure that taxpayers get value for money. The public sector is certainly a major player in the construction market, and local authorities and government departments are therefore encouraged to be aware of the benefits of competitive tendering. Competition in the supply chain is regarded as healthy, whereas arrangements within the supply chain to restrict competition and control prices is unhealthy. Consequently, wherever possible, public (and private) sector clients are encouraged to vet tenders and bids for evidence of collusive agreements. Examples of the types of things to look out for are shown in Table 8.1.

It is understandable that competition needs to be carefully monitored by government and there is a substantial history of **competition policy** that applies right across the economy. For example, in the last ten years there have been investigations into allegations of anti-competitive activity covering industries as diverse as toy retailing, horse racing, construction, newspaper distribution, insurance, crematoria, private schools, bus transportation, groceries, healthcare, mobile phones and airports. Competition policy attempts to restrict unethical business behaviour that acts against the public interest. It is represented by statutory measures operating at the national and European level. In the UK, competition policy is managed by two agencies: the Office of Fair Trading (OFT) and the Competition Commission. Their respective powers as independent enforcers of consumer legislation and competition law are enshrined in the Enterprise Act 2002. For the most recent information on the promotion of competition, readers should visit the respective websites of these organisations which were reviewed on page 30.

As implied above, and explicitly stated in Chapter 2, competitive markets are central to achieving production techniques that do not waste inputs, involve fewer welfare losses and increase growth. Consequently, an effective competition policy that prevents the development of anti-competitive behaviour is important. The closing thought here, however, should remind us of a suggestion made towards the start of this text – namely that, in the final analysis, questions relating to sustainability can be reduced to effective resource allocation – and that is only achievable in a freely competitive market framework.

Key Points 8.5

❑ An efficient allocation of resources is associated with markets that operate freely.

❑ As markets move closer to the perfectly competitive extreme, sustainability policies become more achievable. Hence an efficient system for monitoring free competition has become an important government policy objective.

Two important aspects of Part A have been to explain the importance of resource efficiency and to overview the characteristics of various market structures. We suggested that, in most cases, the market structures in the construction sector are different from those found in the manufacturing sector. Manufacturing is mostly dominated by a concentration of very large companies that are able to utilise capital-intensive modes of mass production and benefit from economies of scale. Whereas, in direct contrast, the construction industry is traditionally characterised by a large number of small firms, with very few barriers to entry, a dispersed market structure and a relatively low level of fixed costs. Which sector achieves greatest resource efficiency (value for money) is left open to debate.

In this extract Blismas, Pasquire and Gibb argue that cost benefit evaluations of off-site production (OSP) are currently too focused on costs, and that they need to become far more holistic to take full advantage of the resource efficient benefits offered by OSP. While studying the extract, therefore, a useful exercise would be to try to identify some of the cost-based evaluations to account for the slow uptake of OSP and some of the broader resource efficient advantages that would make it more economically viable.

Nick Blismas, Christine Pasquire and Alistair Gibb (2006) 'Benefit evaluation for off-site production in construction', *Construction Management and Economics* 24: 121–30

Introduction

Recent UK government reports, including the Egan Report Rethinking Construction (1998), produced by the Construction Task Force, discussed the need for performance improvements in the UK construction industry. Egan (1998) identified supply chain partnerships, standardisation and off-site production (OSP) as having roles in improving construction processes. The Australian construction industry has also recently identified OSP as a key vision for improving the industry over the next decade (Hampson & Brandon, 2004).

The uptake of OSP in construction is limited however, despite the well documented benefits that can be derived from such approaches (Neale et al., 1993; Bottom et al., 1994; BSRIA, 1999; CIRIA, 1999,2000; Housing Forum, 2002; Gibb & Isack, 2003). A major reason posited for the reluctance among clients and contractors to adopt OSP is that they have difficulty ascertaining the benefits that such an approach would add to a project (Pasquire & Gibb, 2002). The use of OSP, by many of those involved in the construction process, is poorly understood (CIRIA, 2000). Some view the approach as too expensive to justify its use, whilst others view OSP as the panacea to the ills of the construction industry's manifold problems (Groak, 1992; Gibb, 2001). Neither of these views is necessarily correct.

The benefits of OSP are largely dependent on project-specific conditions, and the combination of building methods being used on a project. Decisions, regarding the use of OSP, are consequently unclear and complex. Direct comparison of components is not usually possible due to interdependencies between elements, trades and resources. These complexities make the derivation and use of holistic and inclusive evaluation methods difficult. The unlimited combinations of components, site conditions and degrees of OSP do not permit the derivation of a comprehensive evaluation system; however sufficient common factors exist for a degree of valid comparative analysis.

A pilot study by Pasquire and Gibb (2002) demonstrated that decisions to use OSP are still largely based on anecdotal evidence rather than rigorous data, as no formal measurement procedures or strategies are available.

In other words, current evaluation methods used to compare traditional and off-site produced building solutions do not adequately account for 'value' and therfore cannot 'record' the benefits that OSP can promote.

This paper investigates the proposition that current evaluation methods for OSP are cost – and not value – based, and therefore cannot account for the recognised benefits of OSP. The consequence of this is that OSP invariably appears as an expensive alternative to traditional on-site options. The next section identifies the main benefits of OSP from previous research...

Benefits of site production

The benefits attributed to OSP are numerous and well documented. Gibb and Isack (2003) conducted a large interview-based survey in which they determined construction clients' views on the benefits of OSP. Their findings showed that clients' perceived the benefits of OSP as being mainly time – and – quality based. Table 1 (in the next column) summarises their findings in descending order of benefit.

Further, interviewees of Gibb and Isack (2003) were asked to rank a list of key benefits from the initial interviews and litera-ture, noting both the importance of the benefit and the likelihood of realising the benefit (see Table 2).

The interviewees rated benefits in non-direct cost terms, such as minimisation of on-site operations; reduction of site congestion; reduction of on-site duration; improved health and safety, etc. Direct cost benefits did not feature in these ranked lists (Table 2), although identified within Table 1.

These findings clearly demonstrate that, although OSP can offer direct monetary benefit (in terms of costs), the main benefits are from indirect savings (so-called non-cost items).

Table 1 Clients' perceptions of the benefits of OSP (from Gibb & Isack, 2003)

Benefit	Description
Time	Less time on-site—speed of construction*
	Speed of delivery of product
	Less time spent on commissioning
	Guaranteed delivery, more certainty over the programme, reduced management time
Quality	Higher quality—on-site and from factory*
	Product tried and tested in factory
	Greater consistency—more reproducible
	More control of quality, consistent standards
Cost	Lower cost*
	Lower preliminary costs
	Increased certainty, less risk
	Increases added value
	Lower overheads, less on-site damage, less wastage
Productivity	Includes less snagging
	More success at interfaces
	Less site disruption
	Reducing the use of wet trades
	Removing difficult operations off-site
	Products work first time
	Work continues on-site independent of off-site production
People	Fewer people on-site
	People know how to use products
	Lack of skilled labour
	Production off-site is independent of local labour issues
Process	Programme driven centrally
	Simplifies construction process
	Allows systems to be measured

* indicates high incidence

Table 2 Rating of benefits from highest to lowest according to importance and likelihood (from Gibb & Isack, 2003)

Benefit	(from highest rated to lowest rating)	Cost-related
1.	Minimises on-site operations	
2.	Reduces congested work areas and multi-trade interfaces	
3.	Minimises on-site duration	b
4.	Improved health & safety by reduction and better control of site activities	
5.	Produces high quality or very p ctable quality finishes	
6.	Minimises number of site personnel	a
7.	Benefits when only limited, or very expensive on-site labour	
8.	Enables existing business continuity	
9.	Can cope with restricted site storage area	
10.	Enables inspection and control off-site works	a
11.	Provides certainty of project cost outcomes	
12.	Provides certainty of project completion date	
13.	Less environmental impact by reduction and better control of site activities	

a = cost-related; b = impact on cost.

Based on these findings, pure direct cost comparisons will favour traditional on-site operations that are costed on a rate-based system, with overheads, access, cranage, repairs and reworks hidden within preliminary costs. OSP costs are usually presented as all-inclusive amounts with a premium for off-site capital costs. Having established that the benefits of OSP are largely identified as non-cost items, the paper continues to analyse several cases to ascertain the emphasis of current OSP evaluation methods....

References

Bottom, D., Gann, D., Groak, S. & Meikle, J.(1994) *Innovation in Japanese Prefabricated House-Building Industries*, Construction Industry Research and Information Association, London.

BSRIA, compiled by Wilson, D.G., Smith, M.H. and Deal, J.(1999) *Prefabrication & Pre-assembly—Applying the Techniques to Building Engineering Services*, Briefing NoteACT 2/99, Building Services Research and Information Association, Bracknell.

CIRIA, compiled by Gibb, A.G.F., Groak, S., Neale, R.H.and Sparksman, W.G. (1999) *Adding Value to Construction Projects through Standardisation & Pre-assembly in Construction*, Report R176, Construction Industry Research and Information Association, London.

CIRIA, and principal author Gibb, A.G.F. (2000) *Client's Guide and Toolkit for Optimising Standardisation and Pre-assembly in Construction*, Report CP/75, Construction Industry Research and Information Association, London.

Egan, J. (1998) *Rethinking Construction*, The Egan Report, Department of the Environment, Transport and the Regions, London.

Gibb, A.G.F. (2001) Standardisation and pre-assembly—distinguishing myth from reality using case study research. *Construction Management & Economics*, 19, 307–15.

Gibb, A.G.F. and Isack, F. (2003) Re-engineering through pre-assembly: client expectations & drivers. *Building Research and Information*, 31(2), 146–60.

Groak, S. (1992) *The Idea of Building*, E and FN SponRoutledge, London.

Hampson, K. and Brandon, P. (2004) *Construction 2020: A Vision for Australia's Property and Construction Industry*, Cooperative Research Centre for Construction Innovation for Icon.Net Pty, Brisbane, Australia.

Housing Forum (2002) *Homing in on Excellence—A Commentaryon the Use of Off-site Fabrication Methods for the UKHouse Building Industry*, Housing Forum, London.

Neale, R.H., Price, A.D.F. and Sher, W.D. (1993) *Prefabricated Modules in Construction: A Study of CurrentPractice in the United Kingdom*, Chartered Institute of Building, Ascot.

Pasquire, C.L. and Gibb, A.G.F. (2002) Considerations forassessing the benefits of standardisation and pre-assembly in construction. *Journal of Financial Management of Property & Construction*, 7(3), 151–61.

....

Extract information: Edited and adapted from pages 121–3 of original plus relevant references from pages 129–30.

This extract was written nearly 25 years ago and it restates some of the basic microeconomic theory relating to competition. It has been selected, however, on the basis that it very clearly portrays a comparison between the construction industry and the industries supplying building materials such as aggregates, bricks, ready-mixed cement, glass, plaster and plasterboard, iron and steel, and sanitary equipment. As a group, these are referred to as the materials supply industries and they are of interest since in most cases they seem to be closely akin to the processes and organisation of the manufacturing sector. Indeed, these industries continue to provide some of the best examples of uncompetitive market structures. It is this sharp contrast with the nature of markets in construction that makes the paper relevant to this text, especially as the materials supply industries have such a significant impact on the performance of the construction industry. It was estimated at the time by Hillebrandt (1984: 253) that well over 40 per cent of British construction output may have been accounted for by material and component inputs, and this estimate accords with similar studies at an international level (Xiao and Proverbs 2002: 430). Intriguingly, however, Lowe's paper drew the conclusion that the material suppliers did not take advantage of the powerful position of monopoly and oligopoly that theoretically was available.

We thought it would be challenging to try and identify why the potential to collude and take advantage of monopoly power was not exploited at the time and to determine if the situation is any different in today's market. It might also be interesting to briefly reflect (or discuss) the question that has already recurred several times in the text – namely, does economic theory satisfactorily explain the actual nature of competition in the construction industry?

John Lowe (1987) 'Monopoly and the materials supply industries of the UK', *Construction Management and Economics* 5: 57–71

Introduction

In the UK, the market situation of the materials supply industries stands in stark contrast to that of the construction industry. The construction industry is characterised by a very large number of firms, a dispersed market structure, and a low degree of capitalisation and mechanisation. By comparison the materials supply industries tend to be dominated either by a single firm or by a handful of very large firms able to utilise capital intensive modes of production so as to take advantage of the economies of scale offered by the concentrated market structure and the potential for mass production.

Both construction and materials supply can be said to span several markets rather than a single market. In construction the divisions between the various markets – civil engineering, industrial building, speculative housebuilding, etc. – may be best viewed as a series of overlapping contour lines with considerable interaction at the margins. This applies since many construction firms will operate in more than one sector and because labour, capital, and materials are, to some degree, transferable between sectors. By contrast, in materials supply, the divisions are more clear cut and while groups may operate in more than one sector, there will be little interaction other than competition between different materials, e.g. timber versus metal windows.

The purpose of this paper is to analyse the UK materials supply industries from the point of view of the incidence of monopolistic market structures, the reasons why such a situation prevails and the impact of such on the present and future performance of the UK construction industry.

Market forms and market concentration

Before commencing with this analysis some clarification of basic economics is necessary. The four market forms beloved of traditional economic theory – perfect competition, imperfect competition, oligopoly and monopoly – are not easily defined in precise terms. In this context they may be better viewed as sectors of a linear continuum with the two 'ideal types' – perfect competition and monopoly – at either end with an infinite range of possibilities between the two extremes (Fig 1).

Fig 1 Market Forms

```
    X ............. X ............... X ................. X
monopoly oligopoly imperfect    perfect
                   competition competition
```

In consequence of the above, the boundaries between individual markets can be somewhat blurred. While pure monopoly and perfect competition are extremely rare in practice, there are instances when individual markets come close to the ideal.

A complication arises in this context from the point of usage of the term 'monopoly', which has acquired a different meaning in general speech than that employed in economic theory. A monopoly is sometimes defined as a firm having a substantial turnover or a large share of the market in question rather than the strict economic definition of a single firm controlling the whole of that market. Under the terms of anti-monopoly legislation in the UK, action may be taken to prevent a merger or take-over if the resulting conglomerate would control more than 25 per cent of the market in question or have a turnover in excess of a stated figure. Such would be described as oligopolistic in terms of economic analysis.

This paper is concerned with both monopolistic and oligopolistic market situations – the left hand sector of the continuum in Fig. 1. Thus any product where there is either a dominant firm or firms will be covered by this analysis as would a quasi-monopolistic situation created by restrictive trade agreements.

Market concentration, or more specifically, seller concentration is concerned with the degree to which a small number of firms control a significant share of the market – for example, an industry would be described as highly concentrated if 80 per cent of the employment is in the four largest firms (Bannock et al., 1972). This is thus a major determinant of the market form.

A concentrated market can arise through growth of firms or equally through mergers. Merger activity can take the form of horizontal integration, vertical integration, or conglomerate mergers.

References

Bannock, G., Baxter, R.E. and Rees, M. (1972) *A Dictionary of Economics,* Penguin, London, pp. 78–9

Extract information: From pages 57–8 of original plus relevant reference from page 71.

Part B

Protection and Enhancement of the Environment

WEB REVIEWS: Protection and Enhancement of the Environment

On working through Part B, the following websites should prove useful.

www.bre.co.uk

The Building Research Establishment is a research-based consultancy with offices in England and Scotland. It has a particular expertise in the area of sustainable construction and its website advertises its latest products. Up-to-date information on the BREEAM schemes and whole life costing, introduced in Chapter 9, is available at this site.

www.ends.co.uk

ENDS is an environmental data service, providing a daily news service on European environmental affairs. The homepage provides the opportunity to sample the organisation's authoritative monthly report and has links to other environmental resources on the web. The information can be used in conjunction with all the chapter themes in this section – visit and see why it claims to be the best environmental website.

www.foe.co.uk

For a different perspective, it is often interesting to look at information presented by non-government organisations. The Friends of the Earth site, for example, challenges the rise of corporate power. Pressure exerted by Friends of the Earth contributed to the decision by the construction firm Amec to pull out of building the Yusefeli dam in Turkey. Friends of the Earth's website also has extensive links to other governmental and non-governmental sites.

www.buildoffsite.org

As the name implies, Buildoffsite is an organisation formed to promote modern methods of construction. It was established with government backing in 2005 to create a step change in the application of off-site techniques within the construction sector. It was initially set a target of raising expenditure on these techniques tenfold to £20 billion by 2020. Its members are drawn from all sectors of the UK industry, including developers, designers, contractors, manufacturers, clients and government; the current membership list exceeds 60 organisations. The website contains links to events, publications, case studies and a quarterly newsletter. As you will sense from Chapters 7 and 9, this site is of increasing interest.

www.usablebuildings.co.uk

The Usable Buildings Trust was initially funded by the *Building Services Journal* and the UK government. It is now registered as an independent charity that aims to encourage green building design by focusing on performance in use. The trust's first project – the Post-Occupancy Review of Buildings and their Engineering (PROBE) – ran from 1995 to 2002. These initial studies have a separate link on the site. To date the trust has made a detailed post-occupancy review of more than 30 green buildings, details of these can also be accessed from this website. There is also a brief review of 30 books dealing with the theme of 'usability', and some of these we have drawn from in this text.

9 Markets for Green Buildings and Infrastructure

An important consideration for any firm seeking to control the market and stand out from its competition is to satisfy, or create, a niche market – to produce a service or product that is in some way different from its rivals. In economic terms this is referred to as **product differentiation**. We have already discussed how in the extreme case of perfect competition we assume that the market consists of homogeneous products, in which each individual firm in the market produces an identical product (or service) and has a horizontal demand curve. To express it another way, in a perfectly competitive market there is only one specific 'undifferentiated' product (see Key Points 8.1).

Providing a firm can manage to differentiate its product or service from other similar products – even if only slightly – it can gain some control over the price it charges. Firms producing a differentiated product are able to achieve some independence from their competitors in the industry. They should be able to raise their prices, and thereby increase profits, without losing all their customers. Unlike firms operating at the perfectly competitive extreme, they face a slightly downward sloping demand curve. In fact, the greater a firm's success at product differentiation, the greater the firm's pricing options – and the steeper the demand curve.

OPPORTUNITIES TO DIFFERENTIATE CONSTRUCTION PRODUCTS

Economics textbooks usually emphasise that the opportunities to differentiate a product or service in the construction industry are limited. Firms may be able to market themselves as somehow superior to their competitors in terms of quality or reliability, but they are always constrained by the large number of firms that compete and produce close substitutes. Consequently, the ability of one firm to significantly raise its prices above that of its competitors is restricted. Gruneberg and Ive (2000: 92) extend this hypothesis. They argue that the tendering process creates a further complication, as it is usually assumed that all those selected to submit tenders are undifferentiated – equal, in terms of the service they are offering.

An important aim of this chapter, however, is to identify the economic arguments that may encourage construction firms to take up the green challenge. This depends upon firms in the industry taking the opportunity to differentiate their product by moving away from traditional techniques to those that demonstrate environmental awareness. It also involves paying attention to global, local and user concerns if firms are to develop and construct green buildings and infrastructure.

At the time of writing, a construction firm producing environmentally sensitive products would be able to distinguish itself so effectively from the majority that it could secure short-term monopoly profits – that is, until the time when competitors recognise the benefits of following the same mould, bringing the market back to something nearer to perfect competition and, in this case, bringing the market closer to the idea of sustainability. A trend for sustainable construction is slowly emerging and being taken up by some contractors and clients, and traditional specifications are being challenged in favour of those that demonstrate environmental benefits. The common characteristics of environmentally sensitive specifications are discussed in the next section.

Emerging Green Markets

Even in manufacturing – with supply based on factory techniques and where products are demanded and used by a single customer – it is difficult to develop a market for environmentally superior products. In construction the challenge is even more complex, as there are fewer standard prototypes and often the 'users' of construction products are not the owners. As we have suggested in preceding chapters, each construction product can be regarded as unique. Products are assembled on site by a team of subcontractors. The large labour force is often one stage removed from the agreement made between the client and the contractor. And, as a final twist, the interests of the users are often different from those of the investors that produce the original specification. This makes it difficult for those supplying the products to final users to communicate effectively through market signals. Yet it is in the marketplace where people display their green credentials.

It is therefore not surprising that green development in the construction industry has been relatively slower than in manufacturing – but it is emerging. The most activity has been seen in the commercial sector, with owner-occupiers beginning to specify bespoke headquarters that reflect their corporate ethos. The level of green activity within the residential sector, however, is not so evident, as the main companies engaged in house building have been slow to see the market potential of adopting an environmentally aware corporate image. There are some exceptions, with some green developments by housing associations and some examples of architect-designed homes – eco-homes – for environmentally conscious clients. Finally, awareness is emerging in the sector specialising in infrastructure, which could make an important contribution once it takes off. We now look at each of these sectors in turn.

THE COMMERCIAL SECTOR

Each year, the largest amount of new building work is in the commercial sector (see Table 5.2, page 75). Most of this activity continues to be producing a standard undifferentiated product that tends to be over specified, fully air conditioned and energy guzzling. However, an increasingly significant proportion – say 25 per cent – of the new additions are able to boast environmentally friendly features. Sensitivity to the environment is an increasingly important issue, and businesses and organisations want to reflect their environmental credentials in the types of office that they rent and own. There appear to be good corporate arguments in favour of situating offices in buildings that minimise global and local impacts, reduce energy bills and facilitate greater worker productivity.

According to the Building Research Establishment Environmental Assessment Method (BREEAM), and its US equivalent Leadership in Energy and Environmental Design (LEED) developed by the United States Green Building Council (USGBC), it is possible to audit and assess a broad range of issues within the design, procurement and management of an office building. For example, a detailed evaluation can be made of the materials selected and the energy systems employed to light, heat and cool the building. Interestingly, both of these assessment methods identified the new commercial office market as having most potential and this sector became the testing ground for various BREEAM and LEED schemes. The BREEAM scheme for new office designs was launched in 1990, and the LEED equivalent followed eight years later in 1998. Subsequently, schemes to evaluate existing commercial buildings, homes (both new and old) and various other outlets such as shops, schools health centres and industrial units followed.

BREEAM and LEED have the advantage of sharing nearly two decades of experience and their websites now boast that more than 100,000 buildings have been certified in 41 different countries throughout the world. The majority of these, however, are still in the UK and the United States. These figures, however, simply represent the number of environmental assessments that have been carried out by the Building Research Establishment, the United States Green Building Council or their authorised assessors. It would be more interesting and informative to know how many other green buildings exist that have not been put through an environmental assessment scheme. Either way, the number of green buildings is certainly on the increase.

Construction firms seeking to differentiate their products on the basis of their environmental performance need to deploy their assets in a distinctive way. There is a new breed of commercial client emerging that needs to know that their requirements can be competently fulfilled by the contractor. There are a range of features that typify state-of the-art green developments, and the common ones are listed in Table 9.1.

Table 9.1 The characteristics of a green building

✓	Makes maximum use of natural daylight
✓	Minimises consumption of fossil fuels, by techniques such as natural ventilation, combined heat and power, and orientation of site to benefit from passive solar energy
✓	Reduces the use of fresh water by using grey water recycling for landscape irrigation, flushing toilets, etc.
✓	Minimises site impact by careful landscaping and the preservation of local ecosystems
✓	Reduces the quantity of 'virgin' materials used and selects those that have the least negative environmental impact
✓	Reuses and recycles existing buildings and sites
✓	Minimises material waste during construction and demolition

Source: Adapted from Shiers (2000: 354)

Architecture that is based on (some of) the features outlined in Table 9.1 is slowly emerging. Some of these examples of green buildings are listed in Table 9.2. The ones selected in the table are on, or near, a university campus – so you might have the opportunity to take a closer look.

Table 9.2 Examples of green buildings in the UK

These buildings have been developed since the early 1990s. They are listed in chronological order, with the most recently opened building at the foot of the list.

Queens Building (School of Engineering), De Montfort University, Leicester

The Inland Revenue Building, Nottingham

Elizabeth Fry Building, University of East Anglia, Norwich

Learning Resource Centre, Anglia Polytechnic University, Chelmsford

Wessex Water Headquarters, Bath

Architectural and Planning Studios, University of the West of England, Bristol

Peninsula Medical School, Universities of Exeter and Plymouth

The Gherkin, 30 St Mary Axe, London

The National Assembly for Wales, Cardiff

THE RESIDENTIAL SECTOR

The existing stock of houses in the United Kingdom exceeds 26 million units. In recent years, much of the new housing stock has been built on greenfield sites that are car dependent. The majority of these new homes are low density and inefficient in terms of energy usage. In contrast, governments have sought to promote development on brownfield sites designed around good public transport and utilising high-density designs that exceed the minimum expectations for energy efficiency. The government would also prefer to see developments that include provision of social and/or affordable housing. These conflicting priorities highlight the dilemmas that governments face in supporting sustainable construction.

To compound the government's frustration, resource efficient, environmentally friendly housing is by no means 'rocket science' – indeed, technically it can be achieved easily by most contractors. Take energy efficiency as an example: all that is needed is greater levels of insulation, the careful sealing of all joints, the positioning of windows to make the most of sunlight, and use of a heat exchange system where air going out preheats the air coming in. Most volume developers, however, have been reluctant to adopt such energy efficient measures because of the extra cost (and care) involved. For example, in the 1990s, Wimpey, then one of Britain's biggest house builders, shelved its plans to build green homes. Wimpey did not regard energy conservation to be a good selling point, especially at a time when mortgage interest rates were very high (Chevin 1992: 7). However, more than a decade later energy has become relatively expensive and interest rates have fallen, so nowadays the idea of a super-insulated, energy efficient home has become more attractive and, given the interest in conserving the environment, they are no longer completely exceptional. In fact, the type of building that we have described is now collectively referred to as a 'passive house'; a term used to indicate the small amounts of energy that these buildings require for space heating.

Table 9.3 Five principles of sustainable housing

1	Improve thermal efficiency to a point where homes can achieve zero carbon energy usage
2	Reduce mains water consumption by collecting rainwater and recycling grey water
3	Maximise the use of local, reclaimed and recycled materials
4	Promote public transport and car pools to create a lifestyle that is less car dependent
5	Design into the estate services to enable on site composting, home delivery of grocery and recycling

Source: Adapted from Desai and Riddlestone (2002: 20)

There are currently 5000 passive houses built every year in Europe, mainly in Germany and Austria (Kaan and de Boer 2006: 2). In the UK they are still rare, but an excellent example is the Beddington Zero (fossil) Energy Development (BedZED) project. This has provided homes to 82 families since July 2002. There are five principles of sustainable housing that underpin the BedZED development. Most of these are easy to replicate and make economic sense. For example, the properties have walls with 300 mm of insulation (three times the typical amount), incorporate triple glazing, utilise heat exchange units and make good use of south-facing conservatories to achieve reductions in energy requirements to a point where there is no need to install central heating. This represents a saving of around £1,500 per home, which makes the expenditure on the increased insulation more acceptable. In reducing the energy requirements by 90 per cent compared to that required by a typical home which meets the UK building regulations standards, it is possible to power the entire BedZED estate with a small combined heat and power unit running on renewable fuel such as woodchip. Another achievement was sourcing most of the building materials necessary for the buildings from within 35 miles of the site. The BedZED project also integrates office and leisure facilities, built to the same energy efficient standards, on the same site, enabling people to reduce their dependence on cars as they can work, rest and play within a small neighbourhood.

If we are serious about sustainable construction, the development of passive houses and passive commercial buildings similar to those built on the BedZed estate are important. Indeed, the architects of BedZED sometimes boast of achieving Britain's first zero carbon or carbon neutral development. If this type of development becomes commonplace in the property market, as a nation we would become less dependent on fossil fuels which would help the country meet its obligations to reduce carbon dioxide emissions. In fact, an authoritative survey of the literature (combining more than 80 national and regional studies) indicated that there is a potential to reduce global carbon dioxide emissions by approximately 29 per cent by 2020 in the residential and commercial sectors (Urge-Vorsatz et al. 2007: 388). These calculations were drawn upon by the Intergovernmental Panel on Climate Change in its fourth assessment report, where it concluded that the biggest potential saving in any sector (including transport) related to energy generated for use in buildings.

This discussion suggests that firms specialising in house building (or other structures) could benefit by differentiating their product in several ways and by demonstrating a greater awareness of the techniques and specifications that support sustainable construction. In this way, they could win business in the marketplace by beating their rivals at a new game. As the sources behind the IPCC report made clear, achieving a low carbon future is dependent on new programmes and policies for energy efficiency in buildings that go well beyond what is happening today (Urge-Vorsatz et al. 2007: 395).

INFRASTRUCTURE

This sector represents, in value terms, approximately 15 per cent of construction output each year. It encompasses the construction of railways, airports, tunnels, bridges, power stations, coast and river works, and water supply and wastewater treatment facilities. Each of these products can be specified with sustainability in mind. Indeed, the Institution of Civil Engineers presented its first awards to recognise environmental excellence in the summer of 2003. The CEEQUAL (Civil Engineering Environmental Quality Assessment and Award Scheme) is an audit-based assessment similar to the Building Research Establishment Environmental Assessment Method (BREEAM) but appropriate for non-building projects. It shows how infrastructure may be constructed in an environmentally friendly manner. The characteristics that identify green infrastructure are in many ways similar to those listed in Tables 9.1 and 9.3 – the minimisation of waste, use of recycled aggregates, protection of landscape, ecology and archaeology, management of noise, and efficient use of water and energy.

EXISTING BUILDINGS

The major challenge of sustainability in the built environment relates largely to existing buildings, homes and infrastructure. Most estimates suggest that on average only 1 per cent of a nation's buildings are replaced each year. To take housing as an example, even in the most productive years the completion of new housing rarely exceeds more than 220,000 units. However, the government's desire to raise and promote environmental awareness has complicated the housing development process and slowed down the number of proposals being submitted for planning permission. In the 1960s and 1970s, for example, there were usually well in excess of 300,000 new houses completed each year in the UK. Yet during the period 1996 to 2006 the equivalent average number dropped to 190,000 units. In other words, more than a third fewer new houses were being built than in the 1970s despite their being an economic boom. At this rate of new house building, it would take more than 100 years to replace the existing traditional housing stock with new environmentally efficient dwellings.

Key Points 9.1

❑ The development of green buildings is important to sustainable construction.

❑ Product differentiation can lead to short-term monopoly profits.

❑ A construction firm may differentiate its product by introducing environmental specifications, and opportunities to achieve this are slowly emerging in the commercial, residential and infrastructure sectors.

❑ There are increasing numbers of green buildings (see Table 9.2) which display several common characteristics (see Tables 9.1 and 9.3).

RESOURCE EFFICIENCY

Implicit in the characteristics of green buildings and infrastructure is a better use of resources. This is particularly well illustrated by the BedZED project and equivalent passive house developments in Europe, which achieve a 90 per cent reduction in energy resource consumption. Similar levels of resource gains are evident when construction firms reuse and/or recycle materials, develop brownfield sites, minimise waste, promote public transport and employ local labour. Indeed, achieving greater levels of output with fewer resources lies at the very heart of achieving sustainable construction.

Some analysts argue that much greater resource efficiency is achievable. In the 1990s, an important optimistic report – *Factor Four: Doubling Wealth, Halving Resource Use* (Weizsäcker et al. 1998) – claimed that resource productivity could be increased by a factor of four. Obviously such an increase in efficiency would reduce the demands placed on the natural environment. To demonstrate that a quadrupling of resource productivity was technically possible the report included fifty examples. Twenty were related to energy productivity in various contexts, from refrigerators to hypercars; a further twenty were concerned with material productivity, ranging from residential water efficiency to timber-framed building. Finally, there were ten examples of transport productivity, spanning the benefits of videoconferencing and locally produced goods. Encouragingly, in the context of construction economics, more than half of the 50 examples were relevant to the markets for green buildings and infrastructure. Some of these examples are listed in Table 9.4.

In describing the 50 examples, Weizsäcker et al. (1998), highlight the competitive advantages that could be achieved by exploiting resource efficiency. The possibilities and opportunities given in *Factor Four* are achievable by most firms in any part of the world seeking to differentiate their products. In most industries, if producers are offered the opportunity to adapt production to make it significantly quicker, of consistently higher quality and with a fourfold saving of resources, they

Table 9.4 Examples of quadrupling resource productivity

✓	Steel or timber frame versus concrete
✓	Renewable sources of energy in Scandinavian countries
✓	Air conditioning versus passive cooling
✓	Getting the village feeling in the city: urban villages
✓	Renovating old terraced derelict slums
✓	Superwindows and large office retrofits
✓	Photovoltaics at 48 volts DC
✓	Conservation versus demolition

Source: Adapted from Weizsäcker *et al.* (1998)

would give it a try. Construction, however, is notoriously slow to take advantage of any new opportunities that present themselves. The debate concerning the slow uptake of off-site production was introduced in Chapter 7 and Reading 2. This debate has been ongoing for more than twenty years and forms an important part of the sustainable construction agenda. It has been informed and given impetus by several government reports. For example, the Egan report (1998) encouraged the industry to realise the benefits that the controlled environment of a manufacturing plant could offer to achieve reductions in construction costs, delivery times and in building defects; yet ten years later Egan was sad to note that little had improved (Building 2008: 10). Similarly the National Audit Office (2005) highlighted that modern methods of construction made it possible to build up to four times as many homes with the same amount of on-site labour while reducing on-site construction time by up to half. The next section explores some of the arguments for and against the uptake of resource efficient methods in the construction sector.

OFFSITE CONSTRUCTION METHODS

Modern methods of construction utilise a number of innovations that transfer work from the construction site to the factory. They embrace a variety of approaches referred to several different terms, such as off-site manufacturing (OSM), off-site production (OSP), prefabrication, lean construction and modular build. Common examples of building elements that are produced using off-site production techniques are bathroom and toilet pods, and timber and steel frame structures and roofs. The wide range of benefits that follow as result of adopting these methods are summarised in Table 9.5.

Clearly these benefits would assist a firm to achieve resource efficiency, improve product quality and achieve a greater level of profit. Yet, despite the advantages of these modern off-site technologies, several barriers are reported by construction firms, such as higher capital costs, difficulties of achieving significant economies of scale, concerns relating to manufacturing capacity, the fragmented nature of the industry's structure, skills shortages and a risk-averse culture (Pan et al. 2008: 61).

As Weizacker et al. (1998) pointed out, the constraints to achieving gains in

Table 9.5 The benefits of modern methods of construction

✓	Off site working leads to improved safety
✓	Controlled factory environment leads to increased productivity
✓	Running costs are reduced as air tightness and energy efficiency are improved
✓	Reduced on-site construction time
✓	Significant reduction in costs
✓	Less waste from surplus and damaged materials
✓	Fewer defects and fewer environmental impacts

resource productivity are not technological but institutional. Subsequently this finding has been reinforced by a survey of off-site construction methods amongst 100 house builders in the UK, where several firms claimed that a lack of previous experience prevented them from a wider take up of the modern approach (Pan et al. 2008: 62). In other words, inertia and cultural barriers are regarded as underlying problems. This line of argument suggests that reforming the processes of construction, or development generally, and introducing a more sustainable approach is as much a challenge to our personal values as to our political and economic systems.

Capital Costs Versus Running Costs

A significant example of inertia is the way markets tend to favour the short term in preference to the long term. To some extent the example cited by Chevin (1992), in which plans for energy efficient homes were shelved to reduce prices for home buyers in a market characterised by high mortgage rates demonstrates the nature of short-termism, and this problem is particularly common whenever one person pays for the efficiency gains and another party reaps the benefits. This is easy to see in the commercial sector, in which the priorities of landlords and tenants are frequently regarded as distinct. An often-quoted general rule for traditional commercial buildings is that running costs outstrip capital costs by a ratio of 10:1 over a 25 year period. More specifically, a study carried out on behalf of the Royal Academy of Engineering (Evans et al. 1998) estimated that the costs for a typical commercial building over a 20 year period are in the ratio of 1 (for construction costs): 5 (for maintenance costs): 200 (for staff costs). Yet the present culture in the construction industry still tends to place far greater emphasis on the initial capital cost, while demonstrating little regard for the costs incurred by end users. In terms of efficiency, this attitude creates major resource cost implications – indeed the figures suggest that we may be more than ten times better at wasting resources than using them.

This line of analysis creates another opportunity for construction firms to differentiate their products and services. An integrated approach – which fully takes into account the end user – makes it far easier to suggest that the construction firm is adding value to a client's future business. Yet a client seeking to place an order for a business headquarters that makes use of natural materials, sunlight, energy efficiency, low noise, green plants and a genuine feel-good factor for their employees would find its choice of contractors greatly restricted.

PRODUCTIVITY

It is important that office buildings are conducive to work, yet in many cases there is anecdotal evidence to the contrary. There are even accusations that some offices buildings cause employees to suffer headaches, feelings of lethargy, irritability and lack of concentration – and, in some cases, can be responsible for high rates of absenteeism. Even more worrying are the suggestions that the office environment can cause irritation of the eyes, nose, throat and skin. Although some of these symptoms sound like the side effects of spending an evening in the pub, or too long in the swimming pool, they are distinguished by being prevalent among the workforce of some office buildings and not of others. In fact, the symptoms usually disappear after a few hours of leaving the 'affected' building. This type of condition is commonly referred to as **sick building syndrome** (SBS) and it clearly leads to an inefficient use of human resources. It is important to remember that ultimately buildings are 'machines for working in' and investment in green construction should also result in a more efficient working environment.

An interesting example of a highly integrated green building is the Rocky Mountain Institute in western Colorado. Here it is claimed that the staff that work in the building are productive, alert and cheerful all day – without getting sleepy or irritable. Weizsäcker *(1998: 13)* attributes the high rate of productivity to 'the natural light, the healthier indoor air, the low air temperature, high radiant temperature and high humidity (far healthier than hot, dry air); the sound of the waterfall (tuned approximately to the brain's alpha rhythm to be more restful); the lack of mechanical noise, because there are no mechanical systems; the virtual absence of electromagnetic fields; ...the green plants'.

Occasional days of sick leave mean that employees are being paid but not in return for any productivity. Equally worrying, and damaging to overall productivity, is situation where employees do attend work but spend a part of each day complaining about their working environment – and ultimately they might be so fed up that they decide to look for another job. This clearly all adds up to a waste of resources.

The annual cost of absenteeism from the workplace in the UK has been estimated to exceed 1 per cent of GDP each year – that is approximately £9 billion in current prices (Chatterji and Tilley 2002: 669). This figure, however, does not fully account for the effects of sick building syndrome which has received little attention from economists. As the following calculation suggests, this is a significant omission.

According to a survey (Hedges and Wilson 1987) involving employees working across 46 office buildings of varied age, type and quality the incidence of sick building syndrome is quite widespread. Participants of the survey were asked how much they thought the physical conditions of the office influenced their productivity.

The majority thought that their productivity was affected by at least 20 per cent. This is the equivalent of taking one day off in five. Worker self-evaluation, however, may be subject to exaggeration. But even if we take a reduced figure of 10 per cent, this would still represent a significant cost. For example, if we assume that £20,000 is the average office salary, then an organisation employing 1,000 people could be losing in the region of two million pounds each year. (The calculation is simple: a 10 per cent SBS effect on lost productivity represents £2,000 per employee per year, multiplied by 1000 gives a potential loss of £2,000,000.) Arguably, these figures are a worst-case scenario, and not everyone is equally affected by SBS – the literature suggests that 55–60 per cent of staff in problem buildings may be affected. It seems more plausible, perhaps, to accept an estimate of one million pounds per 1000 employees per year. The more worrying statistic is that the service sector employs more than 15 million people everyday in offices. This calculation implies a national cost of SBS in the UK in the region of £15 billion. Again this may be a worst-case scenario, as it assumes that all buildings are affected by SBS. However, the important point is to consider how certain types of construction can result in inefficient uses of resources and set parameters for debate.

As the Egan report (1998: 22–3) stressed, construction needs to be viewed as a much more integrated process paying far more attention to the needs of the end user – even to the extent that completed projects should be assessed for customer satisfaction and the knowledge gained fed back into the industry. To a limited extent this is happening, and the features likely to influence and improve indoor environmental quality and productivity are listed in Table 9.6. The general message is that construction needs to change its approach. The end user requirements need to be given as much respect as the construction specifications. As the National Audit Office (2001: 44) acknowledged, badly designed buildings fail to meet the needs of the end users, and investment into good quality design and construction could result in a more efficient working environment and lower running costs.

Table 9.6 Internal features to improve productivity

Careful attention to a building's specifications can enhance internal environmental quality and improve productivity beyond the levels achieved in buildings which use standard practices. These characteristics are likely to assist the indoor quality.

✓	Natural systems of ventilation
✓	Building materials and furnishings that have low toxicity
✓	Use of natural daylight
✓	Energy efficient lighting with a low flicker rate to reduce headaches
✓	User control of temperature and ventilation
✓	Attention to maintenance and operation of buildings to reduce the build up of microbial agents

Source: Adapted from Heerwagen (2000: 354)

In terms of economics, the important point that emerges is that as the breadth of expertise required from a construction firm increases the number of firms supplying the market decreases; in short, there is greater opportunity to differentiate between firms. We have already pointed out this type of consequence in Chapter 6 when we reviewed PFI contracts and there are interesting parallels. Successfully completed projects, of either the green or PFI variety, have the potential to be more economically efficient and more sustainable. But both project types seem, at present, to favour the big firm and not the small firm that typifies the industry.

One possibility in the longer term is that teams of small firms will begin to work together more closely to secure a place in the green market. This was one of the ways that the Egan report hoped the industry would go forward. As Egan (1998: 32) expressed it: 'Alliances offer the co-operation and continuity needed to enable a team to learn and take a stake in improving the product. A team that does not stay together has no learning capability and no chance of making the incremental improvements that improve efficiency over the long term.'

Key Points 9.2

- ❏ Thinking long term instead of short term makes an important contribution towards achieving greater resource efficiency in the built environment.

- ❏ Some analysts argue that resource productivity can be increased by a factor of four, and that the barriers to achieving these gains are cultural rather than technological.

- ❏ For traditional commercial buildings, the running costs outstrips the capital costs by a ratio of at least 10:1.

- ❏ An important consideration of any economic activity is to consider the end user. For example, in construction, the internal design of an office building should be conducive to work.

LIFE CYCLE ANALYSIS

It should be apparent that any firm interested in producing products for the green market needs to consider a broad range of criteria. And the few firms that have begun to take their environmental performance seriously have adopted auditing procedures that go far beyond narrow financial measures. By auditing how much energy is used and how much waste is generated at each stage of a product's life, producers can increase resource efficiency and reduce the environmental impact of the product. But deciding where to start and where to stop with these environmental analyses is a contentious issue and the boundaries need to be clearly defined. For example, a construction firm could consider energy efficiency, the reuse of building materials, the energy embodied in the manufacture and transport of materials to site and the use of the building throughout its entire life span, etc. In fact, there seems to be ample opportunities to break into many new markets. In an ideal world, the complete 'cradle-to-grave' aspects of a building would be analysed, but this would take a business into making detailed assessments of first, second and third generation impacts. The important message is to identify carefully the quality and specifications of the product to be marketed, before deciding what is the 'cradle' and what is the 'grave' for specific purposes. Such an approach would take a firm on an incremental journey that would make its product differentiation clear and accountable. Figure 9.1 shows a very simplified model of the opportunities that life cycle analysis might offer to construction.

Figure 9.1 Life cycle analysis of buildings and infrastructure

In this simplified model, the environment is the source of fossil fuel and raw material inputs and a sink for waste outputs.

It is evident that, at each stage, the construction process burdens the environment with many costs. At the beginning of the life cycle, a large amount of natural input is needed for the construction phase and, as is well documented, across Europe the construction industry consumes more raw materials than any other industrial sector. During the operational stage, buildings are also responsible for a very significant amount (40–50 per cent) of greenhouse gas emissions, as buildings rely heavily on carbon-based fossil fuel energy for heating, lighting and ventilation. And finally, at all stages up to and including demolition, there is a large amount of associated waste. In fact, it is estimated that the construction industry accounts for 50 per cent of the total waste stream in Europe.

The life cycle analysis of a building is complicated further by fact that there may be several occupiers with different regimes of repair, maintenance and improvements throughout its life span. At all times, however, there is a flow of resources from the natural environment to the constructed product and vice versa, with varying impacts on the environment at different phases. Consequently, no matter how exemplary the initial environmental specification at the construction stage, the overall impact of a building will be dominated by the way in which it is used.

For our purposes, it is important to remember that we are not dealing here with environmental science. This text seeks to introduce economic concepts and:

- compare ideas of mainstream economists with their environmental counterparts
- understand the interrelationships between the economy and the environment.

These concerns form an important focus of the chapters comprising Part B.

Neoclassical Versus Environmental Economics

Mainstream **neoclassical economics** suggests that market forces determine the specific resources allocated to construction. We introduced these ideas in Chapter 3, where we explained how freely adjusting prices provides an efficient signalling system that determines what is made, how it is made and for whom. (Some readers may wish to review Key Point 3.1.) From this perspective, economists can easily account for why energy intensive, man-made substitutes might be used in place of more environmentally friendly products. Using neoclassical analysis, if inputs become scarce, the price rises; this, in turn, creates an incentive for an enterprising person to identify a gap in the market and produce a substitute. These substitutes often depend upon the clever use of technology and, as time goes on, more natural products are replaced (or substituted) by these man-made equivalents. So, for example, the sharp increases in oil prices during recent decades, which are outlined briefly in Chapter 14 as a cause of global inflation, have highlighted the increasing problem of demand for oil outstripping its supply. Indeed, the price of a barrel of oil reached an all time high in 2008. These significant and persistent price hikes push up the price of petrol and heating, and signal a need to substitute energy derived from oil with energy derived from other sources and to utilise developments in technology to improve energy efficiency. The reason we have presented this seemingly stark simple scenario, in which no explicit account is paid to the environment, is to stress that in traditional economic analysis the whole system is self-determining. In neoclassical terms, there is no need to resort to any form of government intervention to achieve a low-carbon economy, as given time the freely operating forces of the market will make it an economic inevitability.

In direct contrast, **environmental economics** does not accept that the ecosystem, or nature, is merely another sector of the economy that can be dealt with by market forces. Environmental economists proceed from the basic premise that there is an extensive level of interdependence between the economy and the environment; and there is no guarantee that either will prosper in the long term unless governments enforce measures that make firms acknowledge the complete life cycle costs arising from their economic activity. Daly (1999: 81) has crudely characterised the ideas of the neoclassical school: 'The economic animal has neither mouth or anus – only a close loop circular gut – the biological version of a perpetual motion machine.' The important concept that Herman Daly and his environmentally conscious contemporaries bring to economics is the greatly undervalued contribution that the environment makes to the economic system. Indeed, the environment provides all the natural resources and raw materials needed to start any process of building or infrastructure, such as land, fuel and water. The environment also provides mechanisms for absorbing the emissions and waste. In short, in this modern view, the economy is viewed as a subsystem of the environment!

In discussions of sustainability the environmental dimensions cannot be ignored, yet traditional mainstream economic textbooks do not refer to life cycle analysis or any of the equivalent auditing systems that measure environmental impacts. The sole reference point is money and the economy is presented as a linear system – similar to that portrayed by Figure 9.1. To correct this misleading picture, environmental economists usually represent the economic linear system within a larger box, or circle, to represent the environment. This type of approach is adopted in Chapter 11. It is used to illustrate that there is an interdependent relationship between the environment and the economy; that the environment provides resource inputs and carries away the waste outputs and cannot be taken for granted. As an example of a representation of the environmental approach see Figure 11.4 (page 183).

Unfortunately, however, the conventional mindset of those presently managing firms in the construction industry mirrors the approach taken by neoclassical economists. For this to be replaced with a genuine sustainable perspective, a commitment to understanding the ideas of environmental economics becomes most important.

It is worth closing this chapter with the observation that both neoclassical and environmental economists share a common belief that consumers and producers express preferences through their willingness to pay. This may appear ironic, but it seems that in the final analysis most economists are preoccupied with expressing everything in monetary value. This suits neoclassical economists whose main point of reference is the trade of material goods and services in markets at specified prices. It is far more problematic for environmental economists who seek to place monetary values on environmental goods and services that are commonly treated as 'free' goods. We shall elaborate on this further in the next chapter and deal specifically with valuation techniques in Chapter 11.

Key Points 9.3

❏ Life cycle analysis involves a detailed study of the impacts of a product from cradle to grave. In the case of buildings and infrastructure, it emphasises the large amount of resources and waste that are involved in the construction process.

❏ Neoclassical economists hold a strong belief that markets steer economies.

❏ Environmental economists emphasise that the economy is dependent on the environment for several functions that cradle-to-grave analysis helps to value.

❏ Environmental economics offers the construction industry a perspective that could help it to secure more sustainable outcomes.

10 Market Failure and Government Intervention

Throughout Part A, we emphasised that the market system allocates resources efficiently. We described how the price mechanism provides an incentive for firms to enter and exit markets in their search for profits, and how each market arrives at equilibrium. Indeed, up until the last chapter, the dominant theme has been that most economic problems can be resolved by allowing the free market to work (see Key Points: 2.1, 3.1, 5.4, 6.1, 7.1, 8.1 and 8.2). For one specific and intriguing example, see the argument put forward by traditional economists relating to the increasing price of oil as a solution to the problem of climate change, rehearsed on page 160.

The market, however, does not always work. There are some circumstances which prevent the price system from achieving productive and allocative efficiency. This seems to be particularly the case for markets involving or impacting on the environment; markets in which goods are not privately managed but commonly owned. In these cases, non-market alternatives need to be considered. One of the most important non-market forces is government, and this chapter reviews the government's role within failing markets. We shall, however, recognise the possibility that governments can also fail to achieve efficient outcomes, and this is discussed at the end of the chapter.

Market failure describes a situation where the forces of supply and demand do not allocate resources efficiently. It may be defined as:

> a marketplace where the unrestricted price system causes too few, or too many, resources to be allocated to a specific economic activity.

As we suggested in Chapter 2, the majority of environmental problems such as polluted seas, devastated forests, extinct species, acid rain and the vaporising ozone layer are examples of market failure.

Economists refer to these type of problems to justify a role for **government intervention**. As Milton Friedman, a publicist of the market system for more than 40 years, consistently emphasised, the existence of a free market does not eliminate the need for government. On the contrary government intervention is essential as a forum for determining the rules of the game, and as a provider of public goods (Friedman 1962).

WHAT CAUSES MARKET FAILURE?

Traditionally economists identify four common causes to explain why markets fail. The first set of explanations usually concerns the promotion of fair competition between firms, as monopolies and other forms of imperfect competition enable large firms to rig markets and keep prices artificially high. These problems have already been considered briefly in Chapter 8. In this part of the text, however, we are concerned with the protection and enhancement of the environment and, therefore, we restrict ourselves to the other three common causes of market failure. These are:

- externalities
- free-rider problems
- asymmetric information.

Externalities

We introduced the conceptual parameters necessary to understand externalities in Chapter 2 (see Key Points 2.4). We contrasted **private costs** and **external costs** – a distinction that helps to explain a broad set of environmental problems. The related analysis represents an important tradition in welfare economics, stretching back to the beginning of the twentieth century.

The idea that economic efficiency should describe a situation in which nobody can be made better off without making somebody else worse off dates back to around 1890 in work by Vilfredo Pareto, an Italian social scientist. According to Pareto, in a truly efficient competitive market all the exchanges that members of the economy are willing to make have to be agreed at fair prices. In such a situation, nobody can benefit unless they take advantage of someone else. There is a general equilibrium. All members of the economy face the true opportunity costs of all their market-driven actions.

In many real markets, however, the price that someone pays for a resource, good or service is frequently higher or lower than the opportunity cost that society as a whole pays for that resource, good or service. In short, it is possible that decisions made by firms and/or consumers in a transaction will affect others not involved in that particular transaction to their benefit or detriment. To put it more simply, in the competitive marketplace, a deal is struck between a buyer and seller to exchange a good or service at an agreed price; but, alongside this two-party activity, there are possible spillovers to third parties – that is, people external to the specific market activity. The spillover benefits and costs to third parties are termed **externalities**.

To clarify the concept further it might help to draw a distinction between the full economic cost and the basic cost of a good or service. The basic cost takes into account all stages of production, which could include extraction, manufacture, transportation, research, development and other business costs such as marketing. In other words, the basic cost covers all the costs that are usually added up to account for a market price. The full economic cost, however, includes all the possible basic costs plus the externalities; or, to put it another way, the full economic costs are the true burdens carried by society in monetary and non monetary terms. In short:

full economic costs = basic costs + externalities

An example of an externality is the pollution of a river, the air or an open public space caused by a construction process. This leads to a general loss of welfare for a community. If this community is not compensated for its loss, then the cost is external to the production process. The construction firm has created a negative externality. In producing a building, the firm has paid for inputs such as land, labour, capital and entrepreneurship, and the price it charges for the finished product reflects all these costs. However, the construction firm has acquired one input – waste disposal into the river, air or open space – for free, by simply taking it. This is, indeed, taking a liberty; the construction firm is not paying for all the resources it is using. Or, looking at this another way, the construction firm is giving away a portion of environmental degradation free with every product.

Any kind of spillover that causes environmental pollution is called a *negative* externality because there are neighbourhood costs such as contaminated water and loss of habitat and associated health issues such as respiratory problems that society at large has to pay. In other words, these community costs are external to the economic transaction between the construction firm and the purchasers of the completed building. An important goal of environmental economists is to close the gap between private costs and external costs. The aim is to make the polluter pay – to make sure that those responsible for causing the pollution are made to pay the costs. This idea of making the polluter pay is discussed later in the chapter. Note, however, that if these costs are to be invoiced in some way we need to know how much to charge which, in effect, means putting a monetary value on the environment – and we shall look at ways of measuring environmental costs in more detail in Chapter 11.

Before leaving the topic here, however, we should acknowledge that not all externalities are negative. The production of a good or service can generate spillover benefits for third parties. In these instances, the market failure is not so problematic. Governments can choose to finance these goods or services that generate positive externalities through subsidies to the private sector – ensuring that companies are rewarded for production of a good or service that, if left to market forces, would be underproduced. A simpler alternative is for a government to take responsibility for the production of the good or service itself. The next section on free riders will confirm the appeal of this approach.

Free-rider Problems

Whenever positive externalities greatly exceed private benefits, the good or service concerned becomes unprofitable in the market context – in effect, some benefits associated with the good or service are allocated for free. For example, if you pay for several lampposts to light the pathway and pavement outside your house, the private benefit (to yourself) would be too small relative to the cost. And the external benefit to your neighbours from this street lighting would be significant, as they would be getting a brighter pathway for free. The problem is that the market system cannot easily supply goods or services that are jointly consumed. For the market to work efficiently a two-party agreement is preferable. If non-paying parties cannot easily be excluded from the benefits of a good of service, we have the problem of the **free rider**. Good examples of this situation are the markets for sewerage services, public open space, paving, street lighting, flood control, drainage, roads, tunnels, bridges and fire-protection services.

Asymmetric Information

Most economic texts identify the problems created by a dominant firm, or a group of colluding firms, as typical causes of market failure. As an example, reflect on the market structures that typify firms in construction and the possible opportunities for them to enter into agreements on joint profits (or at least review Key Points 8.4). In this text we have chosen to emphasise that *any* contractual agreement that is loaded in favour of one party can contribute to market failure. There is a general problem of one-sided information. In Chapter 6 (see page 86), we introduced the idea of asymmetric information.

> A situation in which some of the parties involved in an economic transaction have more information than others is defined as asymmetrical.

Markets may not achieve efficient outcomes when the consumer has to defer to a more informed producer. Let us develop this idea a little further with a simple example. When most consumers go into a music shop to buy a CD, they have enough information to make a rational decision. When they purchase services from a builder, the situation is often very different. In this situation, purchasers know roughly what they want to achieve – but they must rely on the experience and advice of the builder to specify what precisely needs to be done.

This situation – in which one party holds most of the cards – is a common cause of market failure. A new academic approach to market analysis is emerging that focuses on the contractual agreement between the 'principal' – that is, the client – and the 'agent' – the contractor. This focus on the **principal-agent** relationship questions the balance of power between the less informed client and the knowledgeable agent. The debate is around the extent to which the agent acts in the best interests of the client. This analysis of the principal-agent relationship demonstrates how the skills and experience of the agent could lead to a situation in which a trusting client may be misinformed. The initial discussions on principal-agent relationships appeared in health economics: in health contexts, it is clear that the doctor – the agent – has far more medical information than the patient – the principal. Consequently, we are very reliant on doctors to act in our best interests.

Principal-agent analysis can equally be applied in construction contexts – to project managers, engineers and architects. Many large-scale construction projects are technically complex and not easily understood by non-professionals. Although the costs of a mistaken choice may not appear as dire as in medical cases, they are equally difficult to reverse. For example, if the clients or purchasers of a major building development wish to reduce the environmental impact of the construction process, they are completely dependent on the expertise of contractors to achieve these outcomes. It is quite possible that energy usage may not as efficient as it could be or that waste may not be minimised as requested. The hired 'agent' may not always act in the client's best interest, and they might be able to get away with it because of the 'principal's' incomplete knowledge.

Key Points 10.1

❏ Market failure occurs whenever the free forces of supply and demand over-allocate or under-allocate resources to a specific economic activity. Examples seem to be widespread across the environment.

❏ Three reasons for market failure are (a) externalities, (b) the free-rider problem and (c) asymmetric information.

GOVERNMENT INTERVENTION AND MARKET FAILURE

Governments intervene in various ways to correct market failures. Historically, the preference had been for correction via legislation, but increasingly correction is sought by influencing prices and knowledge. Some typical examples are outlined in Table 10.1. This indicates, in a very general way, some of the approaches that are used to tackle different types of market failure. As the press describes it, stick, carrot and tambourine approaches all play a part. To clarify what this means, in the next sections we discuss some of the approaches used by government to resolve each of the three causes of market failure, before going on to consider their effectiveness.

Table 10.1 Government policies to address market failures

Market failure	Government tax	Government spending	Publicity & information	Government regulations
Externalities	Landfill tax Climate change levy Aggregates levy			Water quality legislation
Free-rider problems		Provision of public goods Tax relief for cleaning up contamination		Habitats and species legislation
Asymmetric information			Carbon trust campaigns Home information packs	Building regulations Code for sustainable homes

Source: Adapted from HM Treasury (2002: 23)

Government Taxation

The UK government collects almost £575 billion in taxes each year. The lion's share is obtained through taxes on personal incomes and business profits, and a relatively small amount raised through environmental taxes such as those charged on waste, carbon usage, pollution and so on. However, the system of taxation is evolving from simply raising funds to provide essential public services to altering patterns of private expenditure. As politicians like to say, the burden of taxation is beginning to shift away from 'goods' such as employment towards 'bads' such as pollution and environmental damage. In other words, taxation is being used to determine the allocation of resources by influencing the final market price of a good or service. Taxes that operate through the price mechanism to create an incentive for change may be described as a **market-based instrument**, as theoretically these taxes seek to internalise external costs into the price of a product or activity. Interestingly many of the recent examples of taxes introduced to reduce negative externalities relate to the construction industry.

LANDFILL TAX

The landfill tax was introduced in October 1996. It was imposed to provide an incentive to minimise waste and promote recycling – to internalise the costs to the community of waste going to landfill. Depending on the nature of the waste, the current tax can be as much as £24 per tonne for so-called 'active waste' that gives off emissions, and £2 per tonne for 'inert material' such as concrete bricks and excavated soil that does not biodegrade. This is potentially a significant penalty, although in the case of construction the vast majority of the waste is inert and attracts the lower rate. Currently, slightly more than 100 million tonnes of construction and demolition waste ends up as landfill – of which 16 per cent apparently is material delivered and then thrown away unused. The government has indicated that it will make significant annual increases in the standard rate of landfill tax, and it is expected to double to £48 per tonne by 2010. There is also an ambitious target to half construction waste sent to landfill by 2012 and achieve zero inert waste to landfill in the foreseeable future (HM Government 2008: 48–9).

Although there is some uncertainty regarding the precise amount of construction waste, it is widely accepted that the industry is the biggest producer of waste products. Hopefully the increasing rates and targets relating to landfill will send a clear signal to those working in construction of the need to reduce the external costs associated with the large volume of waste produced, as well as provide an economic incentive to develop recycling.

CLIMATE CHANGE LEVY

The climate change levy commenced in April 2001. It is basically a tax on the business use of energy, and it covers the use of electricity, gas, coal and liquefied petroleum gas (LPG) used by the non-domestic sector. The levy is imposed on each business energy bill according to the amount of kilowatts used. There are differential rates for different energy sources. The levy is nearly three times higher for electricity (£0.0044 per kilowatt hour) than gas or coal (£0.0015 per kilowatt hour). This differential has been introduced because the use of each type of fuel creates different levels of greenhouse gas emissions. The levy has increased energy costs in the commercial sector by 10 to 15 per cent.

The purpose of the climate change levy is to encourage businesses to internalise – that is, pay for – the negative externalities associated with the greenhouse gas emissions that they are responsible for generating. Firms using environmentally friendly energy technologies, such as photovoltaic systems, energy crops and wind energy, or combined heat and power systems are exempt from the levy. Manufacturing, mining and utilities companies have been hit the hardest by the introduction of this levy. Ironically, the impact on the construction industry could be beneficial: the climate change levy encourages businesses to use energy more efficiently, and as all businesses occupy buildings, the expertise of the construction industry could potentially help to make offices and factories more energy efficient and, therefore, reduce energy costs.

AGGREGATES LEVY

The aggregates levy came into effect in April 2002. It is a tax applied to the commercial exploitation of rock, sand and gravel. It applies to imports of aggregate as well as to aggregate extracted in the UK. Exports of aggregate are not subject to the levy. The purpose of the levy is to give businesses operating in the UK an incentive to compare the full costs – including all negative externalities – of using virgin aggregates with alternatives or recycled materials.

To explain it another way, the levy has been established to reduce the noise and scarring of the landscape associated with quarrying. These environmental costs could not continue to be ignored, and the levy is meant to encourage the polluter to pay. The intention is that the construction industry should reduce its demand for primary materials by recycling as much as possible and by reducing waste on site. The immediate benefactors from the removal of these negative externalities would be those communities living close to the quarries. And it is interesting to note that their opinions were sought in the preparatory research that established the initial level of aggregate levy at £1.60 per tonne. By 2008 it had increased to £1.95 per tonne, and in April 2009 it will see further revision to £2.00 per tonne. Finding an exact value for the environmental costs of quarrying will be examined further in the next chapter.

Government Spending

The second area that we identified as a cause of market failure relates to the free-rider problem. The basic problem here is excludability. The benefits of some goods or services – due to their very nature – cannot be excluded from non-payers. Even supporters of the free market, from Adam Smith to Milton Friedman, have recognised that there are a few goods and services that the market mechanism does not supply effectively. These are generally referred to as **public goods**.

In order to explain the precise nature of public goods, it is helpful to begin at the other end of the spectrum and clarify the definition of **private goods**. Indeed, so far in this text, private goods have been at the heart of the analysis. We have mainly discussed the activities of private construction contractors in providing private goods and services. These private goods (and services) are distinguished by two basic principles. One can be termed the **principle of rivalry**. This means that if you use a private good, I cannot use it; and, conversely, if I use a private good, you cannot use it. For example, when I use the services of a plumber, he or she cannot be working at the same time on your water and heating system. We compete for the plumber's services; we are rivals for this resource. The services of plumbers are therefore priced according to our levels of demand and the available supply of their time, and the price system enables plumbers to divide their attention between customers. The other principle that characterises a private good is the **principle of exclusion**. This simply implies that once a good is provided others may be prevented from enjoying equivalent benefits unless they pay. In short, anyone who does not pay for the good or service is excluded. For example, if a road bridge is set up with a tollgate, then the communications link that the particular bridge offers is available only to those who pay. All others are excluded by the price mechanism.

These principles of exclusion or rivalry cannot be applied to *pure* public goods. They are non-excludable and non-rivalrous in their characteristics. National defence, street lighting and overseas representation are standard textbook examples of pure public goods. A distinction is sometimes made between *pure* public goods, which are both non-excludable and non-rivalrous, and *quasi* (near or impure) public goods, which do not have both these characteristics. The major feature of *quasi* public goods is that they are jointly consumed. This means that when one person consumes a good, it does not reduce the amount available for others. It is difficult, therefore, to apply a discriminatory price system. Construction projects, such as bridges and roads, exemplify quasi public goods – especially if a toll system is enforced.

Figure 10.1 A spectrum of economic goods

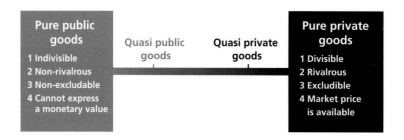

There are four distinguishing characteristics of public goods that set them apart from normal private goods. These four qualities are portrayed in Figure 10.1. This shows a spectrum contrasting the characteristics of pure public goods against those that typify pure private goods.

Developing Figure 10.1, we can describe public goods in more detail as follows.

- Pure public goods are usually indivisible, as these goods cannot be produced or sold in small units.
- Public goods can be used by increasing numbers of people at no additional cost – both the opportunity cost and marginal cost of one more user is normally zero.
- Additional users of public goods do not deprive others of the benefit.
- It is very difficult to charge people for a public good on the basis of how much they use, and they cannot be bought and sold in the marketplace.

TAX RELIEF

Public goods overcome the failure of markets to supply goods or services that generate external benefits. In other words, they enable governments to intervene to provide resources that market forces would otherwise under-allocate. Equally, the government could provide tax incentives or subsidies to encourage the private sector to innovate in a way that will benefit society as a whole, both now and, more importantly, in the future. There are various tax incentives and subsidies to encourage research and development across all sectors. For example, developers are being encouraged to devise ways to clean up contaminated land through the provision of a 150 per cent tax credit for the costs incurred.

REGULATION, PUBLICITY AND INFORMATION

In each of the corrective actions described so far, businesses are being encouraged to reduce the incidence of environmental damage, either by responding to modified price signals that include environmental costs or through the government taking responsibility by providing public goods or paying subsidies through tax incentives. In contrast, another set of options is for governments to set regulatory standards or use their authority to provide information to aid decision-making. As these schemes are relatively less likely to raise business costs, they are considered rather ineffective instruments. But in some instances there are few alternative options, and these mechanisms continue to have a role to play.

GOVERNMENT REGULATIONS

In the history of government intervention in markets there are more examples of regulation than anything else. This generalisation does not only apply to regulations designed to improve markets in the construction sector, but also to the economy as a whole. In this section, therefore, we could address all types of market failure, such as regulations relating to patents and fair trading. However, we shall concentrate on those relating to problems of asymmetric information.

To begin with an example that has many implications across the property and construction sectors, it is the government's responsibility to prevent businesses denying responsibility for the environmental impact of their products and services. This seems to be particularly important as we are increasingly aware that habitat needs to be conserved, water quality needs to be maintained, carbon emissions need to be reduced, sites of scientific interest protected, and building standards observed.

Legislation of this type is well established – for instance, the present system of **planning regulations** has a history of more than 50 years and the set of **building regulations** is more than 20 years old.

According to the Building Act 1984, building regulations can be made in England and Wales for the purposes of securing the health, safety, welfare and convenience of people in and around buildings, to further the conservation of fuel and power, and to prevent waste. Responsibility for complying with the regulations rests with builders and developers. The aim is to assure the public that a certain level of technical accuracy has been achieved in any construction work and environmental impacts reduced.

Building regulations (or codes as they are sometimes referred to) set a baseline of minimum standards to be expected from the industry. As such, they lag behind the standards imposed by the equivalent sets of regulations in many other north European countries. This is particularly the case in terms of energy efficiency, and even though building regulations are continually updated and revised a staggering amount of energy could be saved by going beyond the minimum standards set out in the building regulations. The new code for sustainable homes is a push in that direction as it seeks to improve on the 2006 building regulations, increasing energy efficiency by 25 per cent by 2010, 44 per cent by 2013, and ultimately reaching an energy standard of zero carbon emissions by 2016 (DCLG 2008).

PUBLICITY AND INFORMATION

Information is gathered by the government in an attempt to support and influence market decisions. This approach is particularly important when markets lack transparency due to problems of asymmetric information. Examples take many forms, from the simple government campaign to raise awareness or publicise findings, through to statutory requirements to gather information to support a market transaction. Two contrasting examples of this type of initiative follow.

The Carbon Trust that was established by the government in 2001. This organisation is primarily concerned to raise business understanding of the payback periods involved in moving to a low-carbon economy. It publicises how to reduce emissions and develop commercially viable low-carbon technologies. It boasts that in its first seven years it has worked with 141 local authorities, helping to secure annual savings of more than £70 million and reductions of 861,000 tonnes in carbon dioxide emissions across all sites involved. This represents savings of up to 20 per cent on local authority energy bills. Its website proudly announced its success with an RAF base where switches were replaced with timers, saving £345,000 annually and reducing carbon emissions by 650 tonnes. Not surprisingly, a National Audit Office review published towards the end of 2007 concluded that the Carbon Trust offers value for money in helping businesses reduce their carbon emissions and confirmed that the organisation was destined to deliver an annual reduction of 4.4 million tonnes of carbon dioxide emissions by 2010 (NAO 2007a: 5). Clearly the Carbon Trust is designed to encourage environmentally friendly purchasing and it represents a good example of government trying to help organisations in society avoid inertia and ignorance in regard to environmental prformance. The scheme can be justified as it helps to form a greater symmetry between consumers and suppliers in an effort to create fairer, more transparent and efficient energy markets.

A government can also choose to legislate about the level and quality of information required to support a market transaction. A good example of this type of intervention is the home information pack (HIP). These became compulsory in December 2007. This legislation means that for every home put on the market, no matter what size, the seller must provide a pack of supporting information that includes a sale statement, warranties and guarantees for building work, a condition report, evidence of title deeds from the land registry and an energy performance certificate containing advice on how to cut CO_2 emissions and fuel bills. Prior to the decision to introduce the packs, nearly 30 per cent of transactions in the housing market had collapsed before completion. This represented an estimated waste of £1 million a day on failed transactions. The legislation, however, seeks to make the sale process as transparent as possible from the outset. In other words, the pack guarantees that everyone in the transaction has reasonably complete knowledge about the product as well as potentially raising awareness of questions relating to energy efficiency.

Ideally these campaigns and legislation to improve flows of information should create a greater symmetry between the expectations of consumers and the knowledge of suppliers, leading to a fairer, more efficient market allocation. It is also possible for information measures to reinforce the objectives of new taxes. For example, information on greenhouse gases can help people to respond positively to carbon taxes that differentiate between fuel sources. In this way, improved information services raise awareness and provide more efficient markets.

Key Points 10.2

❑ Governments can use several devices to correct market failure. These include taxes to internalise externalities, the provision of public goods to overcome free-rider problems, and publicity and legislation to reduce the problems associated with imperfect flows of information.

IS GOVERNMENT INTERVENTION EFFECTIVE?

The assumption that the alternative to a failing market is a brilliant government is wrong. Governments can fail, too. Several of the corrective measures we have discussed have problems. These are briefly summarised in Table 10.2 and are examined further in this section.

Measurement Problems

Government attempts to minimise negative externalities require measurement. The **polluter pays principle** is all well and good, providing that the guilty parties are easy to identify and that it is possible to determine a fair price for them to pay. Given that many externalities manifest themselves in global or national environmental issues

Table 10.2 Market failure and government intervention

Cause of market failure	Example of intervention	Nature of problem
Externalities	Taxes and levies	Measurement
Free-rider problem	Public goods Tax credits	Tax burden
Asymmetric information	Publicity campaigns Building regulations	Enforcement

and involve free goods, such as air, the ozone layer, habitat, flora, waterways, and peace and quiet, their measurement (and assessment) causes endless problems.

To analyse these problems further it may help to consider Figure 10.2. Here we have the demand curve D and the supply curve S for product X. The supply curve includes only the private costs (internal to the firm). Left to its own devices, the free market will find its own equilibrium at price P and quantity Q. We shall assume, however, that the production of good X involves externalities that are not accounted for by the private business. These externalities could be air pollution, destruction of a green belt, noise pollution or any neighbourhood cost. We know, therefore, that

Figure 10.2 Internalising external costs

We show the demand and supply for X in the normal way. The supply curve S represents the summation of the private costs, internal to the firm producing X. The curve to the left, S_1, represents the total (social) costs of production. The grey arrows indicate the external costs that have been added. In the uncorrected situation, the equilibrium is Q, P. After imposing a tax ($P_1 - P$), the corrected equilibrium would be Q_1, P_1.

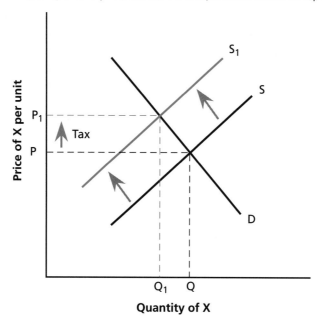

the social costs of producing X exceed the private cost. This can be illustrated by shifting the supply curve to the left, since it indicates that theoretically the costs of producing each unit are higher. (You may remember from Chapter 5 that changes in price and non-price determinants – such as a tax – are represented in different ways in graphical analysis. Review Key Points 5.3 for further clarification.)

The diagram highlights the fact that the costs of production are being paid by two groups. At the lower price P, the firm is only paying for the necessary private inputs. The difference between the lower price P and the higher price P_1 is the amount paid by the community – the external costs. For these external costs to be internalised, the government would need to introduce a tax equal to $P_1 - P$. This should result in fewer resources being allocated to this activity – with less demand and supply Q_1 – as the tax would lead to higher prices and force potential purchasers to take into consideration the costs imposed on others.

It is easy to see, therefore, that in an unfettered market, external costs are not paid for and resources are over-allocated to environmentally damaging production. A tax should help to alleviate the problem, but the practical issues of precisely how much tax and who will be burdened with the expense are difficult questions to resolve.

Tax Burden

As we have explained, pure public goods would not be properly provided by a market structure because of the free-rider problem. For similar reasons, quasi public goods would also be under-allocated. The incentive to contribute to the cost of production of public goods is greatly reduced by the knowledge each individual will potentially benefit regardless of whether they pay. Consequently, most governments step in to provide goods and services such as law and order, overseas representation, infrastructure and environmental management. The concomitant demand for roads, tunnels, bridges, prisons, police and fire stations, overseas embassies, play areas, clean recreational space, flood control systems, etc. explains how governments become such important clients of the construction industry. In the UK, the public sector accounts, in value terms, for about a third of the business done by construction firms. (HM Government 2008: 2).

The drawback to this level of commitment is the cost, especially as the majority of goods that governments produce are provided to the ultimate consumers without direct money charge. Obviously, this does not mean that the cost to society of those goods is zero. It only means that the price 'charged' is zero. The full opportunity cost to society is the value of the resources used in the production of goods provided by the government. For example, though nobody pays directly for each unit of consumption of defence or environmental protection, everybody pays indirectly through the taxes that finances government expenditure.

In the UK, the government collects something approaching £575 billion in taxes each year. Spending on law and order, defence, the environment, international co-operation and transport alone accounts for approximately 25 per cent of this expenditure. In effect, the average citizen in the UK must work from 1 January to March or April just to pay all their direct taxes.

This **tax burden** is clearly a significant proportion of any citizen's income, and it raises some of the thorniest questions that any government has to face. In the UK, and much of the developed world, public spending grew relatively unchecked until the early 1970s, but now many governments choose to exercise some restraint by adopting the **golden rule** – which, in simple terms, means that governments should not allow current spending to exceed current receipts. The golden rule forms a central plank of modern government and its function will be discussed further in Chapter 12.

Enforcement Problems

The success of any government policy cannot rely solely on a strong theoretical argument. Political support, voter appeal and luck are equally important. In other words, just because a government has carefully debated and passed through parliament a new policy, launched a publicity campaign or initiated another set of regulations does not automatically guarantee success.

Rule-based measures, such as regulations, create a whole range of associated costs. There are the compliance costs of implementing, enforcing and administering the legislation. For example, the building regulations are devised to set minimum standards, such as how much insulation should be used, what kind of windows should be fitted and how efficient the heating boiler should be. Yet in recent history no builder has been prosecuted for non-compliance!

In fact, it has been recognised that the 4,000 building control inspectors who work for local authorities are expected to deliver an increasingly complex agenda. A government review entitled *The Future of Building Control* actually acknowledged that 'the system is not broken but it has some serious failings and weaknesses that must be tackled if we are to ensure that it remains fit for purpose in today's world and in the future' (DCLG 2007: 5). Perhaps even more cynically the Conservative party has observed that not only has building control become unfit for purpose, but it is based upon an outdated premise. By prescriptive regulation and complex guidance, it measures inputs and not outputs. It tells builders and architects, surveyors and designers how to do things, not what to achieve (Gummer and Goldsmith 2007: 70).

The Environment Agency experiences a similar set of problems, and it has argued that it has insufficient funds to carry out the number of inspections required by the directives emanating from Westminster and Brussels. For example, in 2002 the Environment Agency complained of a funding gap of £12 million. One of its worries was that the impact of EU directives to phase out landfill sites was leading to an increased amount of fly-tipping.

Even if regulations are enforced, there is little incentive to be innovative. In fact, regulations could be seen to work in reverse and impede change. Some analysts argue that the heavily regulated nature of construction activity creates many of the conservative attitudes that typify the industry. There is usually little incentive to do better than the regulatory standard that has been set, so typically construction firms only do the minimum that is required. For example, few builders will bother to construct a building that exceeds the standards set out in the regulations, unless of course an informed client demands something special. Similarly, if governments rely on publicity campaigns that do not raise the costs of production, then all businesses, construction or otherwise, have no real incentive to refrain from using polluting products and methods.

Businesses can improve performance. Some exceptional examples of potentially resource efficient projects were given in Chapter 9 (see, for example, Table 9.4). Indeed, it was suggested that in some instances it was possible to exceed traditional performance by factors of 4 and even 10. But, in general terms, the construction industry suffers from inertia, and is not particularly innovative or sustainable.

Key Points 10.3

❑ One method for internalising external costs is to impose a tax. But it is difficult to set tax rates so that the polluter pays the correct amount.

❑ To overcome resource allocation problems, governments usually provide a range of public goods. Inevitably, these create a range of associated costs that are ultimately financed by taxpayers.

❑ Just because a government has rubber stamped some regulatory procedures or launched a publicity campaigns does not automatically mean that better practices will be effectively enforced or voluntarily introduced.

GOVERNMENT FAILURE

To conclude this chapter, we should recognise that market failure cannot simply be remedied by government action – that is, perfect governments do not resolve imperfect markets. In fact, modern economic texts also acknowledge the occurrence of **government failure**.

Government failure is understandable, since the political process by its very nature is likely to be inefficient in allocating resources. When choices are expressed through the market mechanism, the price forces individuals to absorb most of the costs and benefits. Politicians, however, allocate resources more on the basis of judgement. Government judgements are often skewed by lack of financial incentives, gaps in information and pressures applied by different interest groups that need to be acknowledged for re-election.

The sheer scale of managing a nation from the centre is problematic. As we have discussed above, there are problems of enforcement, funding and measurement. These problems lead to inefficiency and a wasteful use of resources. Indeed, the more wide reaching and detailed an intervention becomes, the less likely it is that the benefits will justify the costs.

In recent years, therefore, the tendency has been to believe that, in general, markets provide the best means of allocating resources; and this is as true for environmental resources as for others (HM Treasury 2002: 1). Government systems tend to become bureaucratic, inflexible and excessively expensive to run. Furthermore as government intervention increases, individual liberty is reduced and the profit motive declines. The present trend, therefore, is to provide government incentives through the market system. This means that environmental taxes and other economic instruments will continue to be the key tools used to achieve environmental improvements. How far this trend should continue before we reach an 'optimum' level of government intervention is debatable – it is not solely a question of economic efficiency, but one of politics too.

Key Points 10.4

❑ Government failure is a recently acknowledged phenomenon that highlights the fact that constitutional intervention through policy initiatives does not necessarily improve economic efficiency.

❑ Government failure is caused by a number of factors, such as poor judgement, lack of information, inadequate incentives and the sheer scale of the problems to be resolved.

11 Environmental Economics

Economists like to proceed from a model framework to simplify reality and create a reference point for their specific analysis. For many years, however, the models used by mainstream economists tended to overlook the interactions between the economy and the environment. The benefits of economic activity were exaggerated and accounted for, while environmental costs were ignored. We introduced this major distinction when we discussed a product's life cycle (from the 'cradle' to the 'grave') in Chapter 9. However, we have suggested that the environment cannot be ignored since it provides resources at the beginning of a product's life cycle, and absorbs waste at the end of the cycle. This important strand of economic thinking began to gain credence in the 1960s, drawing an important dividing line between environmental and neoclassical economics (see Key Points 9.3).

The significance of the environment for businesses of all types is succinctly represented by the flow diagram in Figure 11.1. The environment is the resource base that provides renewable and non-renewable resources that enable production to begin. At the other end of a product's life, the environment is also expected to provide a sink facility to assimilate the waste matter. (The environment can also provide opportunities, as an amenity, for various leisure pursuits.) The process is easy to exemplify within the construction sector, as the industry consumes resources and generates waste on a scale that completely dwarfs other sectors of the economy.

Figure 11.1 The environment: beginning and end

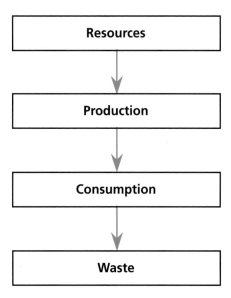

In the first place, it is the environment that provides the land on which buildings and infrastructure are located. Second, it is the environment that provides many of the resources that are used to make building material products. Construction and the operation of buildings account for 40 per cent of the total flow of raw materials into the global economy every year, 25 per cent of virgin wood used and 40 per cent of total energy use (Sustainable Construction Task Group 2002: 2). Finally, but by no means least important, it is also the environment that is ultimately responsible for assimilating and processing the waste that arises from the various phases of construction, from building through to demolition. In other words, without the environment there would be no resources for construction and no way of managing some of the waste and outputs arising from the processes involved in maintaining the building stock and associated infrastructure.

In this chapter, we concentrate specifically on the ways in which the environment and economy interact and explore three important conceptual areas that characterise environmental economics. These are:

- the materials balance model
- private costs versus social costs
- environmental valuation.

THE MATERIALS BALANCE MODEL

The materials balance model has become a standard introductory reference for students new to environmental economics. It focuses on the energy used and wasted by economic activity. The first law of thermodynamics provides the starting point. This law explains the 'physics' of economic activity, since it states that we cannot destroy matter, we can only change it. Therefore, any resource that is extracted from the environment must be returned to it in some form or other – what goes in must come out.

The model conveys the process of production as the transformation of a certain number of inputs into certain a number of outputs. Some of the outputs may be positively valued, but other outputs will be undesired and of negative value. In other words, all economic activity produces a desired output (the conventional good or service) plus undesired output (such as pollution). For example, when clay and sand are combined and heated at high temperatures for several days there are several outputs: the conventional or desired output of bricks, and the undesired output of carbon particles and sulphur dioxide. In short, we have introduced two related notions: the principle of **joint production** that highlights how any economic activity inevitably produces several types of output; and the **materials balance model** that works from the premise that matter cannot be created or destroyed. The concept of joint production is derived from classical economics and the materials balance model from physical science. Their general significance to environmental economists is to emphasise that everything is connected to everything else.

The basic ideas behind the materials balance model and joint production are captured in Figure 11.2. It also illustrates that the mass of material inputs into an economy is balanced by the mass of products and waste outputs leaving the system. In the model, the environment is portrayed as having a similar relationship to the economy as a mother to an unborn child in so far as it provides sustenance and carries away wastes. An alternative perspective is to view the environment as a large protective shell surrounding the economic system. Unfortunately, this environmental shell is often treated as a 'free' good and was for many years ignored by traditional economics.

Figure 11.2 The materials balance model

The material balance for the whole economy is represented by the flows: *A = B+C*

A relevant example of the materials balance process is detailed in the research project *Rocks to Rubble* (NCBS 2003). The researchers accounted for the flow of mineral resources throughout north-west England (including Cumbria, Lancashire, Cheshire, Greater Manchester and Merseyside) during 1999. The aim was to encourage businesses to become resource efficient by thinking beyond the traditional cycle of consumption and disposal. Construction materials provided the focus of the study and it was demonstrated that 31.4 million tonnes of resources produced 21 million tonnes of buildings, roads and other infrastructure, 7 million tonnes of waste and 3.4 million tonnes of recyclable materials. This process of tracking the flow of materials and products through their life cycle provided a clear illustration of the materials balance in practice. It is visually summarised in Figure 11.3, which shows the equivalence between resource inputs and outputs. More importantly, the overriding concerns identified in the study were the inefficient use of resources, pollution and waste. The report (NCBS 2003: 41–2) provided several insights into the necessary changes if society is move towards a more sustainable use of resources. For example, it suggested that waste could potentially be reduced to zero if materials thrown away by one contractor were used by another, though it acknowledged that there seemed to be no means of putting the two together.

This type of study has been used to inform the wider debate on sustainable construction where improvements in materials, technologies, design and waste management are integral to achieving greater resource efficiency. A project cited in *Strategy for Sustainable Construction* (HM Government 2008) offers a far more comprehensive (and complex) study, as it applied a broad definition to the construction industry and referred to the whole of the UK. It estimated that the total mass of all solid waste from the broad construction industry (in 1998) was over

Figure 11.3 Materials balance in north-west England

31.4 million tonnes of
resources

31.4 million tonnes of
product and waste

Source: Adapted from NCBS (2003: 13)

150 million tonnes a year; the lion's share of this waste was from construction and demolition activity (91 million tonnes) with most of the remaining comprising quarry waste. This amounts to nearly 3 tonnes of construction waste for each man, woman and child in the UK, and highlights the scale of the problem in a resource intensive activity such as construction (Smith et al. 2003).

The materials balance model and the related idea of joint production, suggests that maintaining the stock of natural capital is an essential prerequisite for understanding and managing sustainable development. However, according to economists, there are two different approaches to achieving this goal. As outlined in Chapter 9, the basic premise of neoclassical economic analysis is that, left to its own devices, the market mechanism will provide the necessary incentives to encourage technological solutions to resource problems. In short, technology should provide substitutes for any 'shortages' in the environment. As a result, the economy could grow forever. As Robert Solow (1974: 11), a Nobel laureate economist, wrote more than 30 years ago: 'It is very easy to substitute other factors for natural resources...The world can, in effect, get along without natural resources.' An old phrase used to highlight the supremacy of the economic system was that the economy would 'grow around' particular shortages such as resource problems. Technology would invent an easy way out! This traditional technocentric view is based on the **constant capital approach**, in which any decreases in natural resources are substituted by increases in man-made assets.

Environmental economists present things differently. According to their analysis, the economy and the environment are inextricably integrated. The ecosystem and the economic system are viewed as complements rather than substitutes. In fact, most environmental economists proceed from the materials balance model to argue that the economy is a subsystem constrained by the ecosystem, as it depends on the latter as a source of raw materials inputs and a sink for waste outputs. This ecocentric view adopts the **natural capital approach**, which is based on the premise that it is impossible to substitute natural capital with physical capital. According to Lovins et al. (1999: 158), advocates of this type of approach, 'production is increasingly constrained by fish rather than by boats and nets, by forests rather than by chain saws, by fertile topsoil rather than by ploughs'. In other words, the material factors comprising the ecosystem are unique and cannot be substituted at any price.

Empty World Versus Full World

To draw a distinction between these contrasting approaches, it is common for environmental economists to present two models. One, an open system, which we alluded to in Figure 9.1 when we portrayed resources flowing in a linear fashion, that pays no respect to the limits imposed by the environment. And the second, a more constrained closed system, in which the environment physically contains and sustains the economy. Some economists refer to the traditional view of the open system as an 'empty world' model to contrast with the 'full world' model that is the modern closed system.

Figure 11.4 is another way of illustrating the empty world model – the economic subsystem is small relative to the size of the ecosystem. It depicts the past, when the world was empty of people and man-made capital but full of natural capital.

Figure 11.4 Empty world

Source: Adapted from Goodland, Daly and El Serafy (1992)

In contrast Figure 11.5 represents the full world model, describing a situation nearer to today, in which the economic subsystem is very large relative to the ecosystem. This highlights the fact that unless qualitative changes occur the ecosystem is going to be pushed beyond its limits. In fact, there are signs that this point is imminent with, for example, global warming, ozone depletion, soil erosion, biodiversity loss, population explosions and resource depletion.

Professor Kenneth Boulding presented this contrast in a colourful way in his 1966 paper *The economics of the coming spaceship earth*. In this paper he drew an analogy between the empty world and the full world. He referred to the empty world as a cowboy economy, as this effectively characterised the traditional economists' view of the earth's resources – abundant, limitless and capable of sustaining reckless, exploitative and violent behaviour. In stark contrast, he referred to the full world of the future as the 'spaceman economy' – a single spaceship, without unlimited reservoirs of anything, either for extraction or for pollution. The gist of his argument was that the planet should not be pushed beyond its limits. Taking this contrast to its logical conclusion we can envisage two types of economy: an inefficient economy in which all ecosystem services are treated as free goods and used abundantly; and an efficient economic system in which all resources are allocated according to price. This analysis clearly highlights the problems of pursuing unlimited technological growth, and makes a strong case in favour of sustainable growth that does not take the world's natural support functions for granted.

Figure 11.5 Full world

Source: Adapted from Goodland, Daly and El Serafy (1992)

> ## Key Points 11.1
>
> ❑ The economic system and the environmental system are inextricably integrated, since goods and services can not be produced or consumed without the environment providing resources at the beginning of a product's life cycle and absorbing waste at the end of the cycle.
>
> ❑ The materials balance model focuses on the energy used and wasted by economic activity. It prompts a contrast between the empty world of the past and the full world of today.
>
> ❑ There are two contrasting approaches to achieving sustainable development: one based on technology, in which decreases in natural capital are substituted by man-made assets; and another which places far greater emphasis on the critical nature of natural capital.

PRIVATE COSTS VERSUS SOCIAL COSTS

When environmental economists talk about costs they fall into three categories: private, external and social. First, there are the costs of an individual's actions that are known and paid for directly. For example, when a business has to pay wages to workers, it knows exactly what its labour costs are. When it has to buy resources to commence production, it knows what these will cost. Similarly, when tenants have to pay rent for their flat, they know exactly what the cost will be. These are the normal everyday costs associated with most traditional economic activity; they are called **private costs** and formed the focus of Part A. They were formally introduced in Chapter 2 (see Key Points 2.4). Private costs are those borne solely by the individuals who incur them. They are *internal* in the sense that the firm or household must explicitly take account of them.

Second, there are **external costs**, created by the actions of other people. (These have also been discussed previously in Chapters 2 and 10, so it might help to review Key Points 2.4 and 10.1.) We have considered situations in which a business dumps the waste products from its production process into a nearby river, and individuals drop litter on a beach. Obviously, there is a cost involved in each of these actions. As we have noted, when the firm pollutes the water, people downstream suffer the consequences. They may not want to swim in the river or drink the polluted water. In the case of fly tipping or simply common litter, the people who come along after the waste has been dumped are the ones who bear the costs. The costs of these actions are borne by people other than those who commit them. The polluter has not paid, the costs have not been internalised; they are, by result, referred to as external costs.

When we add the external costs to the internal or private costs, we arrive at the third category – total or **social costs**. Pollution problems – indeed, all problems pertaining to the environment – may be viewed as situations in which social costs exceed private costs. Because some economic participants do not pay the full costs of their actions but only their (lower) private costs, their actions are 'socially unacceptable'. If there is a divergence between the social and private costs of a specific activity, we may see 'too many' resources allocated to that activity. To take just one of the many examples, when drivers step into their cars, they certainly don't pay the full social costs of driving. They pay for petrol (which includes a significant element of fuel duty), maintenance, depreciation, road tax and insurance on their cars. However, they cause an additional cost – that of air pollution – which they are not forced to take fully into account when they make the decision to drive. The air pollution created by exhausts is a cost that drivers do not bear directly, but it causes harm to other individuals who suffer the inconvenience of respiratory ailments, dirtier clothes and buildings. The social cost of driving includes all the private costs plus the external costs such as the costs of air pollution, which society bears. Decisions made only on the basis of private cost, therefore, lead to 'too much' driving!

Externalities

As we know from Chapter 10, when a private market cost differs from a social cost it is a problem of **externalities** – individual decision-makers are not internalising all the costs. In other words, some of the costs remain outside of the market and are external to the decision-making process. The individual decision-maker is the firm or the customer, and external costs (and benefits) will not enter into that individual's or firm's decision-making processes. The important point for environmental economists to highlight is that the full cost of using a scarce resource is borne one way or another by society. That is, society must pay the full opportunity cost of any activity that uses scarce resources.

It may help to view the problem as it is was presented in Figure 10.2. There we used a market demand and supply diagram to emphasise that the conventional supply curve includes only internal, or private, costs. Consequently, the market-determined equilibrium price and quantity does not incorporate externalities, as these are not taken into account by the individual producers or purchasers. The quantity produced is 'excessive', and the price is too low because it does not reflect all the costs. In Chapter 10, we considered the possibility of internalising the external cost through some form of taxation to correct the market failure and charge the full social costs of production. This theme is repeated in Figure 11.6.

THE POLLUTER PAYS PRINCIPLE

A firm facing a charge or tax on pollution could respond in one of three ways.

- It could install pollution abatement equipment or change production techniques to reduce the amount of pollution.
- It could reduce pollution-causing activity.
- It could simply choose to pay the price to pollute.

The relative costs and benefits of each option for each polluter will determine which one or which combination will be chosen. Allowing the choice is the efficient way to decide who pollutes and who does not. In principle, each polluter is given the incentive to meet the full social cost of their actions and adjust production accordingly. In theory, there should be less environmental pollution and the price paid by the consumer should increase. This is explained in Figure 11.6.

The pollution tax causes the supply curve S to rise to S_1. The effect of the tax is to increase the price of the good. As a result, the quantity consumed decreases as the cost is higher for the purchaser. But the income received by the producer after tax is paid to the government P_p is lower than it was before the tax was introduced P_e. The environmental cost is therefore shared between the producer and the consumer of the good. Their relative contributions depend on the slope of the supply and demand curves. The more competition there is in the market, the less the consumer will pay.

Figure 11.6 The economic effect of a pollution tax

The supply curve S is based on the costs to the firm producing the good, without any consideration of its environmental costs. If a tax is introduced, the firm's total costs costs increase and the supply curve shifts upwards, by the amount of the tax, to S_1. In the uncorrected situation, in which pollution was being taken for granted, the equilibrium price is P_e and the equilibrium quantity is Q_e. After the tax is introduced, the equilibrium price rises to P_t and the quantity sold falls to Q_t.

IS A UNIFORM TAX APPROPRIATE?

Though taxation offers a way of forcing producers to take account of social costs, it may not be appropriate to levy a *uniform* tax according to the physical quantity of pollution. The same activities do not necessarily have an identical impact everywhere – we must establish the amount of *economic* damage rather than the amount of the physical pollution. A major motorway in London causes much more economic damage than a motorway routed through a less populated area. There are already innumerable demands on the air in London. Millions of people could breathe polluted air and thereby incur health problems, such as sore throats and asthma, which may even lead to premature death. Buildings would become dirtier. A given quantity of pollution, therefore, causes more harm in concentrated urban environments than in less densely populated rural areas. If we are to establish some form of taxation to align private costs with social costs – to force people to internalise externalities and to make the polluter pay – we somehow have to come up with a measure of *economic* costs instead of *physical* quantities. Because the economic cost for the same physical quantity of pollution varies according to factors such as population density, the natural formation of mountains and rivers, so-called optimal taxes on pollution should vary from location to location. (Nonetheless, a uniform tax might make sense given that the costs of administering a variable tax, particularly of ascertaining the actual economic costs, are relatively high.) Either way, the dilemma a government faces is how much should it charge for a 'permissible' amount of pollution and some light will be shone on this difficult question in the next section.

PLANNING GAIN

Before closing this section, it should be noted that not all externalities represent a cost, some are benefits. The principle of **planning gain** provides a classic example. These are agreements in which a developer provides community benefits that extend beyond the marketed output in return for gaining consent for the development to take place. In recent years, these agreements have become a recognised feature of planning negotiations. They are justified on the basis that when planning permission is granted, two things happen. First, the development imposes externalities on the local community, and second, the private developers typically benefit from the fact that their land or property is now of a greater economic value. In some sense, from a developer's perspective, planning gain may be viewed as a sort of 'bribe' or 'sweetener' – offered to accelerate the proposal through the planning process. From the local authority perspective, planning gain, if properly negotiated, can open up opportunities to deliver the right infrastructure to support wider community benefits and enable local plans to take shape. Examples of facilities provided through these arrangements include roads, parks, social housing, schools and health centres.

Because of the protracted nature of the negotiations, these 'deals' often remain low profile. Both parties endeavour to secure their own broad interests and there is a wide range of outcomes. The exchange that is agreed as a result may, from some perspectives, seem underhand or unfair, but it has been encouraged by legislation such as the Town and Country Planning Act 1971. In an attempt to make the negotiations more equitable, the introduction of a **community infrastructure levy** is currently under consideration. In effect this would impose a locally collected development tax. The levy might be calculated as so much per dwelling (such as the £18,000 that is currently being charged on some new housing estates in Milton Keynes) or as so much per area (such as the £14 per square foot charged on new office space in Westminster). These two tariff schemes are prototypes of the type of charges being considered. However if the level is set too high, it might cause some development to become unviable. Each local authority will need to take account of market values in their area but, if they get it right, a planning gain system based on a levy, or tariff, could provide the way to a clearer and quicker basis for securing both private and public benefits and developments.

External benefits cause less of a problem than external costs. Consequently, environmental economists tend to give them less coverage. Conceptually, however, they should not prove too difficult to comprehend, as the symmetry of the first two key points suggests.

Key Points 11.2

❑ Internal (private) costs + external costs = total (social) costs.

❑ Internal (private) benefits + external benefits = total (social) benefits.

❑ Private costs are those explicit costs that are borne directly by consumers and producers when they engage in any resource-using activity.

❑ Social costs include private costs plus any other costs that are external to the decision-maker.

❑ When social costs exceed private costs, externalities exist.

❑ One way to make private costs equal social costs is to internalise the externality by imposing taxes or regulations. In theory, these taxes should be equivalent to the economic damage caused by the activity.

ENVIRONMENTAL VALUATION

As we discussed in this chapter and at several points in Chapter 10, to overcome some of the failings of resource allocation created by the market mechanism it is important to attribute monetary values to externalities. This can be attempted through various methods. Three of the most commonly employed approaches are contingent valuation, the travel cost technique and hedonic pricing. Each of these is briefly outlined below.

CONTINGENT VALUATION

The **contingent valuation method** relies on survey data. People are asked, through a questionnaire or interview, to express the value they attach to a specific environmental asset. For example, in the research leading up to the introduction of the aggregates levy in 2002, some 10,000 people living in communities surrounding quarries were asked how much they would be willing to pay, in the form of increased taxes over a five-year period, for the local quarry to be shut down. The question was based on the assumption that the quarry site would be restored to fit with the natural landscape, and that the quarry workers would be found alternative employment. A further 1,000 people, who did not live in close proximity to a quarry, were asked what they would be willing to pay to shut down a quarry in a National Park such as the Peak District or Yorkshire Dales (HM Treasury 2002: 19).

TRAVEL COST

The **travel cost method** places a monetary value on environmental assets by calculating the cost of getting to recreational facilities such as wildlife reserves, forests and canals. The relevant costs include public transport fares and/or petrol, and travelling time.

HEDONIC PRICING

The **hedonic pricing method** attempts to identify an implicit market value for environmental services by analysing the bundle of characteristics that makes up a product in order to attach a specific value to the environmental element. This process has been described as establishing a 'surrogate market'. The method is easily exemplified in the property market: two properties may be identical except for their location, so the price that people are willing to pay for a house close to a river, woodland, beautiful view or conservation area may be quantified.

The Value of Human Life

The underlying principle employed in all valuation exercises is opportunity cost and this is most clearly illustrated when economists attempt to identify the value of a human life. The conventional method of calculation is to base the value of life on estimates of lost (forgone) contributions to the economy; the stream of lost career earnings. In other words, the opportunity cost of the average victim involved in an accident is based on a wage risk approach or, as some economists have preferred to call it, the human capital approach. In contrast, a more recent method follows the principles of contingent valuation: estimates are based upon an individual's willingness to pay to reduce risks and improve safety or, conversely, the amount that people are willing to accept to compensate for a loss of life.

Government economists throughout the world have adopted these seemingly crude methods to evaluate major infrastructure and transport projects. In resource terms, this enables them to consider a dimension that traditional investment appraisal normally overlooks – namely, the value of lives, lost or saved, due to

Table 11.1 Statistical values of human life

Country	$ 000
Japan	8,280
United States	3,472
France	3,435
Sweden	3,106
United Kingdom	2,281
New Zealand	1,625
South Korea	620

Source: Adapted from Miller (2000: 177)
Note: Base year 1995

accidents. The estimated net benefits to society begin to look very different once the appraisal incorporates a monetary value imputed for life.

Table 11.1 shows the official figures used in different countries during the 1990s to value human life. Interestingly, due to different cultures, different stages of economic development and methods of calculation, governments seem unable to agree on the monetary price that should be allocated. The figures suggest that a Japanese citizen's life is worth far more than any other.

The Precautionary Principle

Ascribing monetary values to any externality is an inexact process; the figures produced are no more than estimates, based on value judgements. It is often argued, however, that it is better to have a rough estimate than completely ignore externalities. In fact, this is the thinking that underpins the precautionary principle that most governments have followed since the Rio Earth Summit of 1992. The Rio declaration defined the **precautionary principle** as follows:

> if there are threats of serious or irreversible damage, the lack of scientific certainty should not be used as a reason for postponing cost-effective measures to prevent environmental degradation.

In short, care should be taken today to alleviate the possibility of problems tomorrow. As a consequence, several externalities have been brought into the financial frame. These range from global issues such as atmospheric pollution to

Table 11.2 Monetary value of global ecosystem services

Ecosystem service	Examples of ecosystem function	Monetary value in 1994 prices (trillion dollars per annum)
Coastal	Processing of nutrients Controls pollution Treats waste Provides for tourism	12.6
Oceans	Generates oxygen and absorbs carbon dioxide Provides habitat for fish	8.4
Wetlands	Stores and retains water Storm protection and flood control	4.9
Forest	Regulates greenhouse gases Prevents soil erosion Provides timber and medicines Opportunities for recreation	4.7
Lakes and rivers	Stores and retains water Flood control Opportunities for fishing	1.7
Cropland and grassland	Provides pollination for plants Pollution control Regulates atmospheric conditions	1.0

Source: Adapted from Costanza *et al.* (1997: 254, 256)

local issues such as the loss of an archaeological site. One comprehensive set of examples is listed in Table 11.2 (on page 191). It is based on a controversial paper published in 1997 and subsequently debated in many journals. An important aim of this paper (Costanza et al. 1997) was to make the case that precautionary action is justified: it showed that given the relative importance of ecosystem 'services', we can no longer risk squandering them. The authors wanted to highlight the enormous economic contribution of ecosystem services and stress that it made no sense to ignore externalities which damage or threaten the global ecosystem. According to their calculations, the total economic value of ecological services was approximately $33 trillion per year (in 1994 prices). This sum was nearly twice the annual value of the global output of goods and services.

Clearly the vast economic systems that have evolved over the centuries have degraded the ecosystem. The value of the two systems needs to be equally respected if sustainability is to become a reality. The work of Costanza highlighted this challenge and began a broader and more accepted focus on placing value on environmental services. This can be seen in the United Nations Millennium Ecosystem Assessment project that run from 2001 to 2005, the EU programme Method*EX* that aimed to harmonise and share data on environmental and health externalities, and the work of the UK government on valuing ecosystems (Defra 2007). In each of these projects, it is recognised that presenting monetary values for ecosystem services to account for the costs associated with their degradation is far more complex than measuring the value of services that are marketed. Indeed it is openly acknowledged that there are significant knowledge gaps in relation to the valuation process, but at least work has begun.

COST-BENEFIT ANALYSIS

When the total costs and benefits of a project cannot be adequately represented by market prices, there is a case to use an alternative method of investment appraisal. Resources that are not allocated through markets, and which therefore do not have a market price, need to be brought into the picture. This is where the technique of **cost-benefit analysis** may be useful.

This method of resource allocation includes externalities as part of the process. It therefore manages to include aspects that most methods of resource allocation omit. In these analyses, external costs and benefits are considered together with internal (private) costs and benefits to provide a fuller picture. Obviously, for this technique to work, all issues need to be expressed in a common denominator for a 'total price' to be calculated. As a result, what may not seem viable in conventional financial accounting terms may appear viable in the broader cost-benefit appraisal.

Cost-benefit analysis is easier in theory than in practice. There are the problems of valuation (as we have discussed above) and there are issues about selecting the externalities. Inevitably, the selection of what is – and what is not – relevant relies on normative decisions based on value judgements. For example, when identifying the external costs of a new underground line in London, the economists conducting the cost-benefit study did not include anything about terrorist attacks on passengers.

Despite the difficulties, economists have carried out many cost-benefit analysis studies to evaluate specific public sector investments. Examples include the construction of the Channel Tunnel, the relocation of Covent Garden and the proposed third runway at London Heathrow airport. In most of these cost-benefit studies, a recurring aspect has been the measurement of time saved expressed in monetary terms. The pricing of an indirect benefit (such as time saved) is as difficult as the pricing of an indirect cost, because market forces fail to determine these prices. In fact, a good way to distinguish between direct (internal) and indirect (external) prices is that the private (direct, internal) cost or benefit will always have a definite monetary value. In practice, some external costs and benefits will be expressed in monetary values in a cost-benefit analysis, and others will not. For example, when appraising new road schemes, government economists usually choose not to put monetary values on traffic noise, visual obstruction, community severance, impact on pedestrians and cyclists, disruption during construction and impacts on climate change. Their excuse is that these costs are uncertain and difficult to pin down.

However, the influential Stern Report (2007) claimed that national governments were committing a serious error in ignoring the future costs of climate change. Nicholas Stern urged them to identify and use a social cost of carbon. The UK government responded by establishing a shadow price for carbon of £25 per tonne. This represents what we are willing to pay now to account for the cost of future damage caused by carbon emissions (Defra 2007: 2).

As Stern (2007: 33) explained, consumption is not just about goods that we might buy from the supermarket, as people also value their health and the environment; and although expressing mortality and environmental quality in monetary terms raises profound difficulties it is essential to avoid global disaster. Slightly further on in the report Stern (2007: 54) adds the dimension of 'time', arguing for the need to assess the flow of costs and benefits over a number of years to effectively compare the welfare and value of future generations with those of today. He strongly supports the case of adopting the convention of expressing calculations in present value terms, using the process of 'discounting' (which is described in the next section). In other words, Stern endorsed the environmental economists approach that the best way to tackle the complex problem of carbon emissions is to ensure that each activity which produces carbon is priced in a way that reflects its true cost to society over time.

The recommendations made by Nicholas Stern were taken on board, to some degree, when the Department for Transport (DfT 2007) conducted a cost-benefit analysis to support its enquiry into the proposal to build a third runway at London Heathrow airport. The DfT's calculations assumed a project with a 70 year life span, and it concluded that the total benefits were an estimated £17.1 billion (at net present values) while the total costs were between £11.9 and £12.7 billion (depending on infrastructure connections). These costs include £4.8 billion to allow for the damage caused by climate change (that is, as much as 40 per cent of the total costs). In short, the total economic benefits of a third runway at London Heathrow would outweigh the total economic costs. To put it bluntly, the project generates a net benefit of between £4.4 and £5.2 billion – or, to put it cynically, around the same amount as the cost allowed for climate change. In interrogating the basis of the calculations (such as the amount of carbon emissions projected and the costs allowed to mitigate their damage), critics have accused the government of producing Alice-in-Wonderland economics (Elkins 2008).

Cost-benefit analysis, therefore, does not provide clear-cut answers to public sector investments. It involves value judgements and (often dubious) estimates which inevitably raise concerns about the reliability of the process. The cost-benefit appraisal offers no more than a framework for government departments to build upon when considering various options. It should be emphasised that the technique is merely a method of *identifying* externalities; it does not automatically control them. However, the point is that while private firms tend to only take account of private costs and benefits, the public sector – concerned with wider economic, environmental and welfare issues – should also consider social costs and benefits.

DISCOUNTING

Another difficulty associated with the pricing of environmental externalities – and with cost-benefit analysis generally – is that governments have to assess the costs and benefits of investment today against the cost of environmental damage that accrues over a number of years in the future. Private entrepreneurs face a similar type of problem in making decisions about capital investments that will secure a stream of revenue and profits in the future. The process used to assist with these decisions is known as **discounting**. This involves calculating the **present value** of flows of money that are expected to arise at some time in the future.

Table 11.3 Present values of a future pound (sterling)

Year	Compounded annual interest rate				
	3%	5%	8%	10%	20%
1	0.971	0.952	0.926	0.909	0.833
2	0.943	0.907	0.857	0.826	0.694
3	0.915	0.864	0.794	0.751	0.578
4	0.889	0.823	0.735	0.683	0.482
5	0.863	0.784	0.681	0.620	0.402
6	0.838	0.746	0.630	0.564	0.335
7	0.813	0.711	0.583	0.513	0.279
8	0.789	0.677	0.540	0.466	0.233
9	0.766	0.645	0.500	0.424	0.194
10	0.744	0.614	0.463	0.385	0.162
15	0.642	0.481	0.315	0.239	0.0649
20	0.554	0.377	0.215	0.148	0.0261
25	0.478	0.295	0.146	0.0923	0.0105
30	0.412	0.231	0.0994	0.0573	0.00421
40	0.307	0.142	0.0460	0.0221	0.000680
50	0.228	0.087	0.0213	0.00852	0.000109

The main reason for making this calculation is that any capital outlay has an opportunity cost: this may be the cost of borrowing or the returns that can be made on an alternative investment. In either case, interest rates can be used to link the present with the future. For example, if you have to pay £110 at the end of a year when you borrow £100, the 10 per cent interest charged gives you a measure of the cost of borrowing. So, the present value (the value today) of £110 to be received in one year is £100 if the market rate of interest is 10 per cent.

The point is that £1 in the pocket today is worth more than £1 in the future because, if you have the £1 today, you could be putting it to work to earn interest. From this, it follows that £1 receivable in the future has to be discounted to find what it is worth today. For instance, at a 10 per cent rate of interest, £1 receivable in three years is worth about £0.75 today. To express it another way, at 10 per cent a year compounded annually, £0.75 will grow to £1 in three years. It is not necessary to work these figures out for yourself since there are plenty of programs available for computers, as well as compound interest tables such as Table 11.3.

In terms of cost-benefit analysis, or any other investment appraisal covering a number of years, it is possible to use discounting techniques to compare the present value of the costs with the present value of the benefits. The problem with this technique is that the present value of any sum of money is dependent on correctly judging three factors: the timing and extent of future costs (and income) and the interest rate. The further into the future a benefit is accrued (or a cost is incurred) or the higher the interest rate used, the lower will be the present value of that benefit (or cost). To take an extreme example, if a nuclear accident occurs in 500 years time, imposing a £10 billion economic cost on the unlucky future generation, it will be reduced to a present value of 25 pence assuming a discount rate of 5 per cent. Likewise with benefits: a tree that needs many years to mature and costs a £100 in current terms would, assuming a discount rate of 5 per cent, have a future value after 50 years of £8.70. Discounting, therefore, provides a system for comparing current costs with future benefits; however, it necessarily downplays the cost of environmental damage for future generations.

Key Points 11.3

❑ Placing monetary value on externalities is problematic. There are three basic methods: contingent valuation, travel cost and hedonic pricing.

❑ Cost-benefit analysis accounts for the monetary values of all internal and external costs and benefits to identify a total (social) price.

❑ Government departments undertaking investment appraisal frequently use cost-benefit analysis.

❑ The problems with cost-benefit analysis are: (a) what to include as 'relevant' externalities; (b) how to quantify the externalities in monetary terms; and, (c) how to effectively 'discount' the criteria to judge all values in today's terms.

Reading 4

Part B has emphasised that the construction industry is significantly resource intensive. It has also demonstrated how economists have broadened their analysis to integrate the material world with the natural world. As a consequence addressing issues relating to the environment and taking opportunities to enhance the image of a firm should be of increasing importance to those involved in modern business.

The following extract reports on current attitudes to sustainability and corporate social responsibility in the construction sector. The data is based on a review of public disclosures (such as annual reports and accounts) provided by major construction companies listed on the London Stock Exchange. The findings suggest that relatively few companies at the time positively embrace the sustainability agenda. This research raises a number of interesting problems that might be worth thinking about. As Milton Friedman (1962) emphatically stated, 'there is one and only one responsibility of business – to use its resources and engage in activities designed to increase its profits'. This raises a significant question: why should a business concern itself with social and environmental responsibilities? So long as it engages in free and fair competition, what's the problem? Second, what kind of examples could a researcher look for as evidence of a firm genuinely demonstrating corporate social responsibility? Finally, what kind of measures might a government adopt to promote sustainable development in the future across the built environment sector as a whole?

Danny Myers (2005) 'A review of construction companies' attitudes to sustainability', *Construction Management and Economics* 23: 781–85

Corporate social responsibility

The main purpose of corporate social responsibility (CSR) reporting is to inform stakeholders of a company's environmental, social and economic performance. Consequently it is often used as a means of measuring a company's progress towards sustainability. The problem for investors and consumers, however, is to differentiate between those who approach sustainability as a PR exercise and those who are genuinely committed; a line needs to be drawn between those who 'green-wash' their products and those who actually 'walk the talk' (Professor Ken Peatrie in CSR, 2004, p. 13).

As the research data suggests, in the construction and building sector there are very few on either side of line. 40% of the companies appear completely traditional and report only financial information. At the other extreme there seems to be a modern cohort who offer a separate report on issues of CSR –this group account for nearly one quarter of the sample group if we include supplementary material from company websites. In between these two extremes are companies that are just beginning to engage in social and environmental issues. They incorporate a limited amount of information into their reports, contributing from as little as 1.4% to as much as 18% of the content. Overall the percentage of material relating to CSR in the sample was found to be less than 5%......

This is not particularly surprising as the conventional goal of any company is to maximise profits; achieving a positive financial bottom line is crucial. CSR, however, does not seek to challenge this convention but to create conditions where social and environmental benefits can simultaneously help to drive the business forward. In John Elkington's (1994) elegant phrase 'triple bottom line accounting' must find resonance with the business community for sustainability to be realised.

A firm that is genuinely and wholly committed to the idea of CSR would use it to grow the business. They would embrace the sustainability challenge and integrate the environmental, social and economic dimensions. They would be 'connected and responsive to shareholders, suppliers, communities and customers' (Henriques and Richardson, 2004, p. xix). In short, they would innovate and think beyond the current paradigm. This is what Professor Peattie (CSR, 2004) referred to as 'walking the talk'. In the research sample there was only one company that appeared to be any where near this model – it was Vinci.

Vinci is a huge French multinational company with 2500 business centres, employing 127,000 employees across 80 countries. It is a major player in energy, infrastructure and construction and they work 250,000 sites per year. The chairman's message in the annual report closes by emphasising that: 'Vinci understands its corporate responsibility as…endeavouring to combine economic, social and environmental performance for the longer-term benefit of the community, the company, its employees and shareholders as well as its customers and partners' (Vinci, 2004, p.3). This is subsequently substantiated by numerous examples of best practice.

Vinci is one of three French companies listed in the sample and it is apparent that each of them addresses CSR issues with marginally more enthusiasm than some of their equivalents in the UK. (The cynic might draw attention to the new company law in France that makes a requirement of all plc companies to include a range of social, environmental and labour information in their annual reports.)

Two other companies that formed part of the research sample, Weinberger AG and Dunham-Bush plc, are also not based in the UK. In other words five of the companies cited in the sample are European (this small control group is referred to in the conclusion).

Conclusion

It is important to emphasise that the survey data referred to above is based on publicly available information from websites and the Financial Times Annual Report Service. This enabled an audit of 42 construction majors listed on the London Stock Exchange. The largest part of the industry, however, is made up of small firms. The DTI register lists more than 170,000 private construction contractors working in the UK – 93% of them employ less than 8 people (DTI, 2004). The typical small private firm will have incomplete records and far less grand corporate strategies. The government will be lucky if they report fully on their financial position, let alone their social and environmental progress.

The lack of respect for sustainability within the construction industry would also appear to be a problem across Europe. As pointed out above, five of the companies listed are European and as such offer a small control group, but no significant differences are identifiable. Furthermore, the French and German construction sectors are equally fragmented, with large numbers of small private firms operating throughout (Carassus, 2004). The research findings, therefore, present a typical picture across Europe.

On the basis of publicly available information, and related literature, the survey suggests that some of the exceptionally large companies in the construction sector are beginning to acknowledge sustainability. They recognise that a business is no longer judged solely on the economic value added by a company's activity; it is also judged on the social and environmental value they add (or destroy). Examples of best practice to minimise the negative impacts and develop the obvious business opportunities are stated in their reports. However, it also clear that many companies, particularly the small unlisted ones, have a long way to go, before they can effectively manage the opportunities that emerge from the sustainability agenda.

Compounding this problem is the diverse nature of the industry both in terms of its associated processes and the related

professions. In fact, the Pearce report argued
that before the industry can proceed towards
contributing to sustainable development it
needs to adopt a more holistic definition that
includes property and its management. As
sustainability is concerned with the complete
life cycle, a building needs to be considered
from its design and construction, through the
operational stage, to its deconstruction. To
make meaningful recommendations, therefore, 'it is necessary to understand the nature
of the industry – its size and structure'
(Pearce, 2003, p. ix).

Consequently, even though the construction industry has been given its own sustainability agenda there is nothing to suggest
that it has engaged more efficiently than any
other industry. In fact, the little evidence
there is would suggest the contrary.
Furthermore, the complex and fragmented
nature of the sector will restrict it from
making a rapid transition – if any at all.

References

Carassus, J. (2004) *The Construction Sector System Approach: An International Framework* (CIB Publication 293) International Council for Research and Innovation in Building and Construction: Rotterdam.

CSR (2004) *Corporate Social Responsibility Magazine*—a supplement of Sustain: Built Environment Matters, McClelland Publishing: Manchester.

DTI (2004) *Construction Statistics Annual*, 2004 Edition, TSO: London.

Elkington, J. (1994) Towards the sustainable corporation: win–win–win business strategies for sustainable development. *California Management Review*, 36 (2).

Henriques, A. and Richardson, J. (2004) *The Triple Bottom Line: Does it All Add Up?* Earthscan: London.

Pearce, D. (2003) *The Social and Economic Value of Construction: The Construction Industry's Contribution to Sustainable Development* nCRISP: London, available from: www.ncrisp.org.uk

Vinci (2004) *Annual Report 2003*, Vinci: Paris.

Extract information: From pages 783–4 plus relevant references from page 784–5.

Part C

Economic Growth that Meets the Needs of Everyone

WEB REVIEWS: Economic Growth that Meets the Needs of Everyone

On working through Part C, the following websites should prove useful.

www.hm-treasury.gov.uk/

This was one of the first government websites, and it is still one of the best. The site opens with opportunities to search for information on Budgets, the economy (including forecasts and the latest indicators) and the euro. There is a comprehensive index, which offers an alphabetical listing of around 100 subjects.

www.bankofengland.co.uk

The role of the Bank of England's monetary policy committee and the significance of a central bank is discussed in Chapter 12. For both of these topics the Bank of England website is a useful reference point. The homepage has a comprehensive index and its own search engine. You can easily find the latest monetary policy committee minutes, up-to-date monetary and financial statistics, and summaries of the Bank's most recent *Quarterly Bulletin* and *Inflation Report* – both of which can be downloaded in full (in pdf format). There is also a link to some museum pages that provide a historical description of the UK's role as a financial centre.

www.worldbank.org

The World Bank is not a bank in the common sense of the word. It is owned by 185 member countries with the primary objective of providing financial and technical assistance to developing countries around the world. As a consequence, it collates a broad range of development indicators ranging from GDP to levels of literacy for low-income as well as high-income countries. This is a good website to find comparative macroeconomic data that is updated on an annual cycle and is supported by commentary and methodological notes. For example, the statistics in Table 13.1 could be updated from this site.

www.nao.org.uk

The National Audit Office investigates the services provided by government funding to assure that taxpayers are getting value for money. It produces around 60 reports per year, and makes electronic versions freely available for download from this website. Many are relevant to this text as they review specific public sector projects as well as exploring broad themes such as PFI and modernising construction. Some examples are listed in the references and the website can be searched for more resources by title or text subject.

www.euroconstruct.org

This website specialises in analysing and forecasting constructions markets across Europe. It provides up-to-date information on 19 countries in western and central Europe. The data is provided by research organisations based within each country and there are links to equivalent organisations in the Baltic States and Japan. It is a good source to access comparative data on GDP and construction as well as specific country reports.

Managing the Macroeconomy

The management of macroeconomic issues appears to dominate the pages of the national press. In the 1930s economists and politicians of the industrial world wrestled with the **great depression**, which lead to millions being made unemployed. Forty years later, during the 1970s, the big problem was **stagflation**, a combination of stagnation (slow or zero growth) and inflation (rising prices). Now in 2008, nearly another forty years later, the macroeconomic issue is the **credit crunch**. This has disturbed the long-standing equilibrium in the financial markets and caused fears of rising inflation and recession. Against this background, governments strive to deliver economic stability or, as it is sometimes expressed by politicians in the UK, to achieve high levels of growth that meets the needs of everyone.

In this chapter, we review the policies and objectives that underpin this vision of economic growth that meets the needs of everyone. We shall also consider economic forecasting. All this has direct relevance for anyone who needs to understand, manage and/or plan construction capacity over a medium-to-long time period, as the demand for construction products is always derived from activity in other sectors. Inevitably, then, construction economists need the ability and confidence to interpret economic statistics relating to the wider economy. Economic statistics are considered throughout this chapter and the next, so Chapters 12 and 13 should be read as a related pair.

FIVE MACROECONOMIC OBJECTIVES

All governments, regardless of their political persuasion, seek to achieve economic goals. In economics, these goals are referred to as **macroeconomic objectives**. There seems to be some political and economic consensus about the five dominant macroeconomics objectives: price stability, full employment, a sustained rate of economic growth, a positive trade balance with overseas partners and effective protection of the environment. Each of these objectives is considered in turn below. Recent macroeconomic statistics for the UK economy, which show the extent to which these objectives are being achieved, are presented in Table 12.1 (see page 205).

Stable Prices

Stable prices are crucial for business confidence, facilitating contracts and enabling the exchange rate system to function smoothly; in contrast, persistently rising prices cause problems for most sectors of an economy. Price stability has become the primary objective of most governments that wish to secure long-term growth and full employment. Economists no longer believe that tolerating higher rates of inflation can lead to higher employment or output over the long term. In the UK, the government's target is to keep inflation within a range of 1–3 per cent and the

consumer price index (CPI) is monitored on a monthly basis. A sample of annual CPI statistics for the UK economy is presented in Table 12.1. As the table shows, the general trend has been one of declining inflation, and this is regarded as encouraging, not as an end in itself but due to its economic significance in meeting all other government objectives. Price stability is so central to understanding modern macroeconomic management that we present a full explanation of its measurement and its impacts upon business expectations in Chapter 14.

Full Employment

Full employment does not mean that everybody of working age is employed, as in any dynamic economy some unemployment is unavoidable. For example, there will always be individuals moving in and out of employment, as they change from one job to another, and there will always be seasonal, technological and overseas factors that cause fluctuations in the jobs available in different sectors. Problems arise, however, when there are large numbers of unemployed for long periods of time; as a large pool of unemployed labour represents wasted resources. It also has many costs, not just in terms of loss of output but also in terms of human suffering and loss of dignity. All governments record the number of workers without a job, although the precise way this is measured changes from time to time. At present 'official' unemployment in the United Kingdom is estimated by the number of people registering for unemployment benefit – known as **claimant unemployment**. Unemployment is either expressed as a percentage rate – the number of claimants as a percentage of the total workforce of 28 million – or as an absolute number. As Table 12.1 shows, the unemployment rate has been below 10 per cent for over a decade, with unemployment reaching a high of approximately 2.3 million in the mid 1990s. The figures in Table 12.1 also suggest that lower levels of unemployment are associated with declining rates of inflation.

Sustained Economic Growth

A long-term objective of all governments is to achieve steady increases in productive capacity. Governments measure **economic growth** by the annual change in the rate of output, and the commonly used measure of economic output is GDP – **gross** (total) **domestic** (home) **product** (output). GDP figures are used worldwide as a proxy for a country's progress towards prosperity: since the more money a country makes, the higher its GDP growth, the assumption is that increases in GDP mean that the citizens of that country are enjoying a higher standard of living. The way GDP measures output can be seen as a giant till ringing up all the transactions taking place inside a country. To accurately portray the rate of change of actual output, GDP must be corrected for price changes from one time period to another. When this is done, we get what is called 'real' GDP. So a more formal measure of economic growth can be defined as the rate of change in real GDP over time (usually one year). As the footnotes indicate, the growth data in Table 12.1 has been corrected accordingly. It is, therefore, a clear indicator of **boom** or **recession** and as Table 12.1 shows it is rare for GDP to be negative. In fact, a full recession has not been experienced since 1991. A fuller coverage of GDP and how it is calculated is given in Chapter 13.

External Balance

All international economic transactions are recorded in a country's **balance of payments** statistics. The ideal situation represents a position in which, over a number of years, a nation spends and invests abroad no more than other nations spend or invest in it. Economic transactions with other nations can occur on many levels and, for accounting purposes, these transactions are often grouped into three categories: current account, capital account and financial account. Of these three, the most widely quoted is the current account. This involves all transactions relating to the exchange of visible goods (such as manufactured items, which would include building materials), the exchange of invisible services (such as overseas work undertaken by consultants) and investment earnings (such as profits from abroad). Clearly, in any one year, one nation's balance of payments deficit is another nation's balance of payments surplus – ultimately, however, this is not sustainable and, in the long run, debts must be paid. The data in Table 12.1 show a worrying trend, in so far as the UK current account figures are all negative amounts. However, in addition to buying and selling goods and services in the world market, it is also possible to buy and sell financial assets and these are recorded separately in the financial account. The UK's annual position on its financial account is usually in balance. A further qualifying remark regarding foreign trade is to recognise that balances of payment figures are notoriously difficult to record accurately. (In fact, of all the statistics shown in Table 12.1, the balance of payments estimates are subject, by far, to the biggest amendments.) In practice, therefore, statistics relating to the external balance need to be considered in a broader historical context.

Table 12.1 UK macroeconomic statistics

There are various sources for this data, we have used the Office for National Statistics website (see the ONS web review on page 30). Three points need to be emphasised, economic data is subject to revision, it is calculated in different ways by individual nations, and proper comprehension relies on footnotes.

	1990	1995	2000	2005	2007
Inflation [1]	7.0	2.6	0.8	2.1	2.3
Unemployment [2]	1.7	2.3	1.0	0.9	0.9
Economic growth [3]	0.8	2.8	3.1	2.7	2.5
Balance of payments [4]	−22.3	−8.5	−24.8	−31.0	−57.8

Notes: 1 Retail prices (percentage increase on previous year)
2 Total unemployment (annual average, in millions)
3 Annual percentage increase in real GDP
4 Current account (total for whole year, £ billions)

Protection of the Environment

The environment has gained a high political profile in recent years, and there is a recognition that a healthy economy depends upon a healthy resource base. We have already reiterated several times that the protection of the environment, and indeed its enhancement, forms an important strand of the sustainable development agenda. We have also examined (particularly in Part B) the problem that – left to their own devices – markets cannot deal effectively with environmental impacts on third parties. At present, there is a broad-ranging debate arising from concerns that some economic activity damages the environment, and politicians are beginning to consider environmental protection together with other macroeconomic goals. As a result, governments are increasingly choosing to intervene to influence resource allocation, preserve biodiversity and reduce pollution. However, there is no agreed way of measuring the impact of environmental protection initiatives and so there is no indicator for this objective in Table 12.1. We have already discussed some of the problems of integrating environmental costs into the marketplace and we will briefly assess the implications again when considering the measurement of national output in Chapter 13 and when introducing the possibility of measuring a nation's ecological footprint.

PRIORITIES: AN HISTORICAL PERSPECTIVE

The order of priority that these five macroeconomic objectives are given varies from government to government. But all governments, in all nations, ultimately seek these objectives in their quest for macroeconomic stability. Indeed, since the end of the Second World War, there has been a consensus that governments should take action to stabilise economic activity. For example, the White Paper on Employment published in May 1944 stated that the government accepts responsibility for the maintenance of high and stable levels of growth and employment, and these themes dominated government agendas in the twentieth century.

Since the 1980s, however, the order of priority has changed and governments have become more concerned with curbing inflation. Budget statements made during the 1990s typically emphasised that price stability was considered a precondition to secure high and stable levels of growth and employment. As the political and economic scene evolves, it is foreseeable that at some time in the not too distant future there will be another shift in emphasis, as the objective of protecting the environment becomes sufficiently important to raise its profile and demote inflation, employment, trade and growth from their current positions as governments' top four macroeconomic objectives. In other words, while there may be some doubts at present about the priority to attach to environmental issues, the interest expressed in climate change and sustainable development is gathering momentum throughout the world and a new order is imminent.

Key Points 12.1

❏ To achieve economic stability, five main macroeconomic objectives are pursued: (a) full employment, (b) stable prices, (c) external balance, (d) steady growth and (e) environmental protection.

❏ The order of priority accorded to these macroeconomic objectives varies from government to government.

GOVERNMENT POLICY INSTRUMENTS

In their attempts to achieve their macroeconomic objectives, all governments, regardless of political persuasion, employ the same types of policy instrument. Again, it is only the emphasis that seems to change. These instruments can be grouped into three broad policy categories:

- fiscal policy
- monetary policy
- direct policy.

Fiscal Policy

In the UK, fiscal policy emanates, on the government's behalf, from HM Treasury. **Fiscal policy** consists largely of taxation (of all forms) and government spending (of all forms). The word fiscal is derived from the Latin for 'state purse' – and this is most appropriate as taxation is the main source of income from which governments finance public spending. In short, fiscal policy is concerned with the flow of government money in and out of the exchequer.

Important elements of the current fiscal framework are to make sure that both sides of the government balance sheet are managed efficiently. Any public sector debt must be held at a prudent and stable level in relation to GDP, and borrowing is only acceptable to cover capital expenditure. In technical terms, these two Treasury rules are:

- the **golden rule**, which states that over the economic cycle the government can only borrow to invest and not to fund current spending
- the **sustainable investment rule**, which states that over the economic cycle public sector debt expressed as a proportion of GDP must be held at a stable and prudent level.

Spending that provides goods and services that are to be consumed in the same year as purchase (such as expenditure on drugs in the NHS) is classed as current spending. Spending that produces a stream of goods and services for use over several years (such as funding a new hospital) is classed as capital spending and can be funded within reasonable parameters by public sector borrowing.

The two Treasury rules provide a benchmark against which the government can judge its fiscal performance. They also establish a stable framework for the broader economy, and remove much of the instability created by previous fiscal regimes. This credible and trustworthy approach to fiscal policy has earned the UK an international reputation for managing its fiscal affairs in ways that are transparent, stable, fair and efficient (International Monetary Fund 1998).

Monetary Policy

Monetary policy is implemented in most countries by a **central bank**, such as the Bundesbank in Germany, the Federal Reserve in the USA and the Bank of England. In the UK, prior to 1997, monetary policy was set by the government – in other words, the Bank of England simply followed government instructions. In May 1997, the government established a new monetary policy framework, transferring operational responsibility to an independent **monetary policy committee** (MPC). The committee is responsible for setting interest rates each month to meet the government's overall inflation target. This inflation target is confirmed in the Budget statement each year. At present, the target is for a 2 per cent increase in the annual consumer price index. (The measurement of inflation is explained in Chapter 14.)

The monetary policy committee consists of ten experts drawn from outside and inside government circles. Five members are from the Bank of England (the governor, two deputy governors and two other bank officials), four 'outside' members are appointed for their expertise, and there is one representative from the Treasury who is allowed to participate in the debate but has no vote. The committee has a very specific and important role and it is governed by an organised and accountable process. At its monthly meetings, the panel of experts carry out in-depth analysis of a wide-ranging set of data. For example, recently published monetary policy committee minutes (extracted from the Bank of England's monthly *Inflation Report*) suggest that the analysis includes the general state of the world economy, trends in domestic demand, the labour market, the housing market and the financial markets, and last, but by no means least, various measures of inflation and costs in specific sectors of the economy. The committee works from the premise that interest rates represent a general cost of activity and, therefore, after allowing a period of time to transmit through the economy, interest rates ultimately control the level of prices and aggregate demand. (We shall briefly extend this explanation in Chapter 14.)

To enable the Bank of England to concentrate on issues relating to interest rates, responsibility for supervising individual financial institutions was handed over to a newly created **Financial Services Authority** in 1997. This separating of functions survived a decade of economic stability (from 1998 to 2007), however problems in the banking sector caused by the credit crunch led to calls for the Bank of England to have regulatory powers for the financial system in addition to control of monetary policy. In 2008, the government announced plans to establish a new financial stability committee to run along similar lines to the MPC.

CO-ORDINATION OF FISCAL AND MONETARY POLICY

An important point to note at this juncture is that fiscal and monetary policies are equally important in managing the macroeconomy. A change to either policy has broad effects on many of the core macroeconomic objectives. Consequently, all governments employ both fiscal and monetary instruments, although the emphasis alters from government to government. Until 1997, the Chancellor of the Exchequer directed the operation of both UK fiscal and monetary policy. Although this theoretically meant that there could be a high degree of co-ordination between both arms of macro policy, in practice this was often not the case. The new Labour government took pride in the levels of stability that followed the introduction of its new monetary regime in 1997, and certainly the change resulted in a greater clarity of roles and responsibilities between the Bank of England and the Treasury.

Direct Policy

Many other government economic policies tend to be more 'objective specific' compared with the broad macro fiscal and monetary policy options we have considered so far. We refer to these instruments as **direct policy**, but it is also known as direct control or direct intervention. A feature of this type of policy is that it tends to have less impact on overall market prices than the broad macro changes to tax or interest rates.

Direct policy tends to be of a legislative nature. Conventional economic textbook examples include legislation designed to control prices, wages or imports to assist with the stabilisation of prices and trade; legislation to support research and development, education and training to influence long-run growth; and general support to encourage small businesses. Good examples of direct policy within the area of construction economics include building regulations and codes established to raise the quality and performance of finished products, and the *Strategy for Sustainable Construction* (HM Government 2008) introduced to change cultural attitudes towards productivity, safety and the environment. These initiatives are aimed at stimulating growth, stability and environmental performance within the construction sector.

MACROECONOMIC OBJECTIVES AND POLICY

Effective macroeconomic management is not an easy task. The basic objectives and policies are summarised in Figure 12.1 (see page 210). The scales are used to imply that there are trade-offs to be made between policy and objectives. Trade-offs, however, also occur between one objective and another, and one policy and another.

Let us briefly consider one scenario of macroeconomic instability as an example. Suppose interest rates are increased as part of an attempt to reduce spending and prevent further inflationary pressure that could lead to price rises. As a result, consumer spending is cut back – an action which may reduce employment opportunities. An increase in unemployment would put a strain on fiscal policy as the unemployed would no longer pay income tax (reducing government revenues) and they would receive benefits from the state (increasing government spending). The **public sector net cash requirement** (PSNCR) – the government's 'overdraft' – could increase. In turn, this may necessitate government spending cuts in other areas, leading to further unemployment. As output falls, then obviously economic growth slows down. Yet, as economic growth declines, there arguably would be less environmental degradation. This brief example of the effect of a policy intervention:

- highlights the complex nature of macroeconomic management
- emphasises the potential incompatibility of some macroeconomic objectives and instruments.

Economists would claim that this scenario is now unlikely – arguing that they have learnt lessons from past experience – but they would not dispute the difficulty of the task. As Treasury advisers Balls and O'Donnell (2002: 2) noted: 'Fiscal policy is complicated as there are multiple objectives and multiple instruments.' This difficulty is compounded by the fact that monetary decisions are taken monthly, but fiscal decisions are effectively only made annually in the government's Budget.

Figure 12.1 Government objectives and government policy

Objectives	Policies
Stable Prices	Fiscal Policy
Full Employment	Monetary Policy
Sustained Growth	Direct Policy
External Balance	
Protect the Environment	

Key Points 12.2

- ❏ All governments utilise a combination of fiscal policy, monetary policy and direct policy.
- ❏ Fiscal policy is concerned with government expenditure and taxation.
- ❏ Monetary policy is concerned with achieving price stability through setting interest rates.
- ❏ Effective management of the macro economy is difficult, especially as some of the objectives are incompatible.

MACROECONOMIC MANAGEMENT

Most economies aim to increase their rate of output each year. This has been achieved in the United Kingdom – on average, during the last 50 years, there has been a 2.5 per cent annual increase in economic activity. Unfortunately, however, long-term growth is not achieved at a steady rate, and there are always periodic fluctuations above and below the general upward trend – this concept is portrayed in Figure 12.2. These fluctuations are related to activity in the broader economy.

Figure 12.2 Business fluctuations

The orange line represents the long-term growth trend around which economic activity fluctuates.

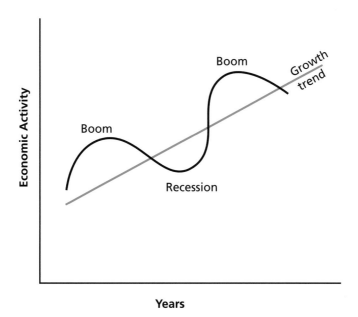

At times, the overall business climate is buoyant: few workers are unemployed, productivity is increasing and not many firms are going bust. At other times, however, business is not so good: there are many unemployed workers, cutbacks in production are occurring and a significant number of firms are in receivership. These ups and downs in economy-wide activity used to be called 'business cycles', but the term no longer seems appropriate because *cycles* implies predetermined or automatic recurrence and, today, we are not experiencing automatic recurrent cycles. The subtle changes that characterise the general nature of economic activity in the twenty-first century are now more usually referred to as **business fluctuations**.

In fact, empirical evidence highlights how the nature of contractions and expansions has changed. A review of data from 28 advanced economies covering the past century shows that the recession phase has tended to shorten in duration, and recessions now rarely exceed 15 months. Periods of expansion have increased, and in some instances in the newly industrialised Asian economies have achieved sustained growth for periods of more than 20 years. In recent years, periods of expansion have become the norm while recessions are relatively rare (International Monetary Fund 2007: Chapter 5). Indeed, strictly speaking, neither the United States nor the United Kingdom experienced a recession (that is, two consecutive quarters of negative growth) during the 10 year period from 1995 to 2005; instead, they had a decade of uninterrupted growth. Although real declines in GDP are no longer common, it is possible to identify periods when growth slows down and is less than the long-run trend. Some experts consider these declines in output sufficient to be classified as a period of recession, and it is interesting to note that the National Bureau of Economic Research in the US now defines a recession as a period of *low* or negative growth in GDP. So the language, terms and character of macroeconomic fluctuations are changing. However, although some journalists and politicians have claimed that the era of boom and bust business cycles is over, this assertion will be tested by the significant economic pressures caused by the (2007-08) credit crunch and spiralling oil prices.

Interpreting Business Fluctuations

There is a collection of economic statistics that can indicate where we are, where we've been and, most importantly, where we seem to be going. Their origin date back to the 1960s and they are still referred to as **cyclical indicators.** They enable governments to predict changes that are happening in an economy. These predictions are based on a composite set of statistics that are regarded as running ahead of the general economic trend. This is because things do not happen simultaneously – some indicators may point in an upward direction while others portray a downward trend especially at the 'peaks' and 'troughs' (the turning point) of a cycle.

Statistics that are assumed to precede the general trend of the economy by changing six-to-twelve months ahead of the main trend are referred to as **leading indicators.** This group is broken down into two subgroups: a longer leading index (which looks for turning points about one year ahead) and a shorter leading index (which indicates turning points approximately six months ahead). Examples of leading indicators are housing starts, new car sales, business optimism and the amount of consumer credit. **Lagging indicators,** by contrast, alter in retrospect, usually about one year after a change in the economic cycle – they confirm what we already know and, in forecasting terms, are not so important. Examples of lagging indicators include unemployment, investment in building, plant and machinery, levels of stock, and orders for engineering output. Economic statistics that are thought to trace the actual cycle are called **coincident indicators** and the obvious example is GDP figures.

Economic Forecasting

The interpretation of economic events is a complex process, especially as macroeconomic policy instruments can affect several variables at once. To take the simplest example, some cynical forecasters are quick to suggest that the sight of an increasing number of cranes on the skyline, visible from their office window, means that we are about to witness the start of the next recession. But this would be a ridiculous suggestion as there are many other variables that need to be used as a basis for forecasting. Indeed the art of forecasting involves completing a picture using as much existing data as possible and combining this analysis with anecdotal evidence to arrive at an overall view. To avoid too many subjective judgements, models tend to be computer based, using mathematical equations to link a number of economic variables. For our purposes, these variables can be categorised into two main types. **Exogenous variables** are *external* to the economy in so far as they are determined by world events and policy (examples include oil prices and exchange rates). **Endogenous variables** are dependent on what goes on within an economy (examples include employment and inflation). There are more than 120 exogenous variables, and hundreds of endogenous ones – and the larger traditional models of the macroeconomy contained upwards of a thousand relationships. The present trend, however, is for models to be smaller and rarely exceed twenty core equations.

Treasury forecasts are informed by more than 40 independent models managed by institutions in the city and beyond. In Table 12.2, we present the May 2008 forecast for the UK economy up until 2012. The table summarises the averages, it does not show the range of comparisons making up the forecast. For example, the highest and lowest forecasts for the percentage change of GDP in 2009 were 2.7 per cent and 0.5 per cent respectively; in 2012, the respective figures were 2.9 per cent and 1.9 per cent.

Table 12.2 Economy forecasts 2008 to 2012

Average projections	2008	2009	2010	2011	2012
GDP growth (%)	1.8	1.8	2.5	2.6	2.5
CPI	2.7	2.2	2.2	2.2	2.1
RPIX	3.5	2.6	2.6	2.7	2.6
Unemployment (mn)	0.84	0.93	0.99	1.00	0.96
Current account (£bn)	-52.2	-46.9	-45.1	-44.8	-44.8

Source: Adapted from HM Treasury 2008

Understandably, since it is difficult to predict accurately the behaviour of millions of consumers and businesses to the last detail, economic forecasts are often wrong. Furthermore, forecasts are limited, since they rely on assumptions about policies that may need to change owing to sudden events or revised statistics. There are also problems relating to time lags, since it often takes years for a specific monetary or fiscal instrument to fully work through an economic system.

The important point, however, is the message conveyed by the forecast; the trend does not have to be 100 per cent accurate. Forecasting models are no different from any other economic model in that they attempt to simplify reality. In the case of the economy, this is a complex reality and a forecasting model only identifies, measures and monitors the key variables. Understanding half of the picture, however, is better than not seeing any of it at all.

Managing the Construction Industry

As you might expect, fluctuations in construction output share a similar pattern to the broader economy and they are referred to as **building cycles.** Interestingly, economists have a long history of studying building cycles, though not because they are particularly interested in the construction industry but because there is a strong possibility that it may contribute to a better understanding of business fluctuations. The symmetry between business fluctuations and building cycles is, however, complicated by the fact that economic development is usually associated with a shift from investment in new construction to spending on repair and maintenance. In other words, as GDP increases the proportion of new construction work decreases. Bon and Crosthwaite (2000) confirmed this pattern. Their international review also demonstrated that there is an inevitable decline in the share of construction in GDP as economies mature. In fact, it has been observed that newly developing countries experience up to double the rate of expenditure on construction as their more developed counterparts. We shall explore this further in the Chapter 13.

Another notable comparison between the construction sector and the rest of the economy is that building cycles have shown far greater amplitude than the equivalent cycles in general business activity. In other words, periods of decline are far more severe in construction than in the general economy, and conversely periods of expansion are more bouyant. For example, on average between 1981 and 2005, a 1 per cent increase in GDP growth was accompanied by a 2.3 per cent increase in total construction output growth (Deloitte 2006: 30). There is, therefore, a unique and distinctive situation in construction – the long-term trend is for growth in construction output to slow as economies mature, yet the variation from year to year can be quite volatile.

This line of analysis is not meant to portray a depressed industry, as construction is, without doubt, a permanent and important sector of any economy. Indeed, in Chapter 13, we will argue that it is possibly the most important sector. It is certainly not like other sectors that may expand and contract and then disappear. The construction sector always has work – to maintain existing stock and to replace demolished stock and construct new stock – even if it may be at a declining rate.

From the perspective of managing the economy, the construction industry is important and most developed economies have a government department or commission to co-ordinate activities in the sector. In the United Kingdom, there are many government interactions with the industry. The departments concerned with roads, housing, health, education, energy, defence and the environment procure products from the industry – in fact, in one way or another, as much as 30 per cent of construction activity, is derived directly or indirectly, from the public sector. There is also a Construction Sector Unit based in the Department for Business Enterprise and Regulatory Reform (BERR) which aims to:

- increase the productivity and competitiveness of the industry
- provide effective links between the industry and government
- encourage the industry to contribute to sustainable development.

In effect, this means that the industry has representatives at government level to speak on its behalf and the government has a mechanism to deliver its messages from the top down. The main day-to-day functions of the unit are summarised in Table 12.3.

Table 12.3 Functions of the Construction Sector Unit

✓ Improving business performance and profitability in construction

✓ Improving the construction process, technologies and techniques

✓ Tackling people issues, such as recruitment, health & safety, training & diversity

✓ Promoting and sponsoring research and development

✓ Improving awareness of the benefits of information technology

✓ Leading on sustainability in construction

✓ Promoting overseas activities by the construction industry

✓ Engaging the industry in regulation and policy development

Source: Adapted from BERR website (2008)

Key Points 12.3

❏ Most modern economies achieve long-term growth, but the pattern is not steady as there are characteristic periods of fluctuation above and below the general upward trend (see Figure 12.2).

❏ Leading (cyclical) indicators are of particular significance in macroeconomic forecasting as they change six-to-twelve months ahead of the main business trend.

❏ Models used for forecasting and policy evaluation are based on computer programs. In the past, these have comprised upwards of a thousand equations linking endogenous and exogenous variables.

❏ The symmetry between business fluctuations and building cycles is complicated. Construction output declines as a percentage of GDP as economies mature, yet the variation from one year to the next can be quite volatile.

❏ Most developed economies have a government department or commission to co-ordinate activities in the construction sector.

The construction industry is an important focus of government policy. This is largely due to the recognition of the importance of construction to national economies. Broadly defined, the construction industry – including manufacturers of building products, equipment and components, and the various professional services provided by architects, surveyors, engineers and property managers – typically accounts for about 15–16 per cent of total annual economic activity. (You may remember that we compared broad and narrow definitions of construction activity in Chapter 1; and it may be here useful to review Key Points 1.3.)

Official statistics, however, generally tend to restrict the construction sector to the narrower definition of the industry, estimating the activity of firms that construct and maintain buildings and infrastructure – that is, just those businesses that undertake on-site activities. Consequently, the share of total annual economic activity attributed to construction by the official statistics in the fully industrialised countries is now rarely larger than 12 per cent and usually in the 7–10 per cent range. In the 27 countries of the European Union, the construction sector accounted for 10.4 per cent of economic activity on average in 2006. Note though, as we explained in Chapter 12 (see Key Points 12.3), that construction output tends to decline as a proportion of a country's GDP as its economy matures.

Apart from the industry's contribution to the total economic flow, it also has a significant impact on living standards and on the capability of society to produce other goods and services. In other words, construction is important to the economy because it produces investment goods. These are products that are not wanted for their own sake, but on account of the goods and services that they can create. Across the 27 member states of the European Union, construction counts for 50.5 per cent of all investment goods. Construction also can have extremely significant effects on the level of employment as it tends to be more labour intensive than other sectors. In fact, in Europe construction is the largest industrial employer, representing nearly 15 million jobs. (All statistics are taken from the European Construction Industry Federation, FIEC 2007.)

In this chapter we explore the relationships between the construction industry, other sectors and the national economy. As the contrast between the broad and narrow definitions of the industry illustrates, there are many sectors closely associated with construction activity. A case can also be made that construction indirectly affects and supports activities in the financial, manufacturing, wholesale, retail, residential and service sectors. Consequently, data for construction-related activities is frequently muddled into manufacturing and service industry surveys. These direct and indirect relationships have important implications for management of the macroeconomy and their analysis is facilitated by the annual publication of **national accounts**.

MEASURING ECONOMIC ACTIVITY

The national accounting framework provides a systematic and detailed description of the UK economy and, by following agreed international accounting conventions, it enables comparisons to be made with other countries. It is not our intention to delve into the minutiae of this system, but just to establish the general measurement concepts necessary to discuss the broader role of construction.

To begin the analysis, we consider a simple economy without a government sector, a financial sector or an overseas sector – that is, our starting point is a simple two-sector model economy and we analyse only the relationship between households and businesses. The complications of the real world will be considered later. We have already portrayed economic activity using this type of a model in Chapter 1 (see Figure 1.4), and for convenience a modified version is presented in Figure 13.1.

To make our starting model effective, we make these assumptions:

- households receive income by selling whatever factors of production they own
- businesses sell their entire output immediately to households without building up any stocks
- households spend their entire income on the output of the businesses.

These three assumptions seem realistic. Businesses will only make what they can sell. Production does involve paying for land, labour, capital and enterprise, and these services generate income payments – rent, wages, interest and profit – which, in turn, are spent. The model of the circular flow outlined in this way suggests that there is a close relationship between income, output and expenditure. These relationships are presented in a traditional format in Figure 13.1.

From Figure 13.1, it is clear that businesses reward the owners of factors of production (land, labour, capital and enterprise) by paying them rent, wages, interest and profit and, in turn, these factor rewards (incomes) form the basis of consumer expenditure. This model shows that it is possible to measure the amount of economic activity during a specified time period by adding up the value of total output, or total income, or total expenditure. In effect, it is only necessary to adopt one of these three approaches since conceptually they are identical – and even in the actual national accounts they rarely differ by more than 0.5 per cent. The small discrepancy is due to each of the totals being calculated using different statistical methods.

To get a better idea of the magnitude of the numbers involved, readers are advised to look at a copy of the *UK National Accounts* (ONS 2007). Table 1.1 and Table 1.2 of the UK edition summarise all three methods of measurement – namely the **output approach**, the **expenditure approach** and the **income approach**. By analysing these statistics, it is possible to gain a good insight into the UK economy, especially as the data covers the last 15 years. Although very few people actually study the detailed breakdown of the accounts from cover to cover, they are an essential data source for anyone concerned with macroeconomics. Indeed, the national accounts are far more important than just indicating changes in GDP; they form a central reference for those who wish to broaden their understanding of the economy and its measurement.

Figure 13.1 The circular flow of income, output and expenditure

The diagram highlights two flows: a monetary flow and a real flow of goods and services. The two lower flows indicate the factor market – households exchange their factors of production with businesses in return for payment. The two upper flows show the product market – businesses provide a flow of goods and services in return for monetary expenditure.

GDP and Growth

Before considering any figures, however, we must fully understand what they convey and the significance of any changes in their size. In simple terms, **gross domestic product** (GDP) can be regarded as the annual domestic turnover; or, to employ the analogy used in Chapter 12, the result of a giant till ringing up all the transactions that occur within a specific territory. In formal terms:

> GDP represents the total money value of all the production that has taken place inside a specific territory during one year.

An alternative measure is **gross national income** (GNI). This is very similar to GDP, but includes a net figure for employment, property and entrepreneurial income flowing in and out of a nation's economy from overseas – in other words, GNI aggregates all the activity that generates income to a specific nationality. In practice, GDP and GNI represent very similar amounts. For example, in 2006 GDP in the UK totalled £1,300 billion and GNI was £17 billion more at £1,317 billion. In European states, GDP and GNI rarely differ by more than 1 or 2 per cent, but the difference may be substantially larger in less developed economies.

When GDP figures are adjusted from current prices to constant prices, to allow for inflation, it is possible to calculate the **real value** of any change in economic activity between one year and the next. Effectively, economic growth can only be declared if 'real' GDP has increased. If real GDP has declined, this is described as a **recession**. In the majority of years during the last half-century the recorded figures have been positive.

Since the UK shares a common set of accounting conventions with other countries, we can make international comparisons of GDP and GNI. Some figures are shown for six selected countries in Table 13.1. The final column, which shows GDP growth, is obviously expressed relative to economic activity in the previous years. The term 'real' emphasises that inflation has been removed from the calculation, with each year's GDP values being expressed at an agreed base year (to convert current prices to constant prices). Worldwide economic activity tends to be on an upward path and, as Table 13.1 shows, all the selected economies grew strongly in the period 2000–2006. In particular, there were big increases in India and China, although both countries were growing from a relatively low economic base. To facilitate international comparison it is necessary to take into account the size of the country. This is achieved by expressing GDP or GNI on a **per capita** basis, by dividing total GDP or GNI by the total population to arrive at an amount per head (see the dollars per capita column in Table 13.1).

Table 13.1 Macroeconomic statistics for selected economies

Countries in rank order	Gross National Income (GNI)		GDP
	Billions of dollars 2006	Dollars per capita 2006	% real growth rate 2000 – 2006
Norway	308.9	66,530	3.3
United States	13,446.0	44,970	2.8
United Kingdom	2,425.2	40,180	2.4
Japan	4,900.0	38,410	1.6
China	2,641.6	2,010	9.8
India	906.5	820	7.4

Source: Adapted from World Bank (2008: Tables 1 and 4)

The statistics in Table 13.1 are taken from World Development Indicators, a comprehensive set of data produced by the World Bank each year. The first table in the series always addresses 'The Size of the Economy'. The current publication lists relevant data for most major national economies in alphabetical order. In previous years, however, this data was presented in rank order according to GNI per capita. The concept of 'rank order' demonstrates the importance of these figures, as they are used to create a type of league table, in which (in 2006) Norway and the United States come ahead of the United Kingdom, since the gross national income (GNI) divided by the population is higher in each of these countries than in the UK. If 'total' GNI were the reference point, then Norway would move to the bottom of our list and the richest nation in the world would be the United States with a GNI in 2006 of $13,446 billion. As you can see from Table 13.1, there is a huge difference between the largest and second largest economy (measured in terms of GNI per capita). There are 299 million Americans and 5 million Norwegians and, once population size is taken into consideration, Norway can offer a higher material standard of living than the United States – $66,530 compared to $44,970 respectively. At the other extreme, China and India with the highest populations in the world are relatively poor, with an annual per capita income of $2,010 and $820 respectively. According to the World Bank, low-income countries are those with less than $905 per capita (which is probably less than some students in the Western world earn each month – even while they are studying) and this gives some idea of the poor average standard of living in India.

GDP and Construction

Construction is a significant part of the total economy. In 2006, construction in the narrowest sense of the definition produced about 7 per cent of UK GDP. In comparison, manufacturing produced 18 per cent of GDP, while agriculture accounted for just 1 per cent. The lion's share of economic activity fell into the service category, which broadly defined includes the wholesale and retail sectors. In 2006, the service sector accounted for more than 74 per cent of GDP (ONS 2007).

In newly developing countries, the construction sector can contribute as much as 20 per cent to GDP because it accounts for a significant amount of investment during a country's development. As industrialisation proceeds, factories, offices, infrastructure and houses are required and construction output as a percentage of GDP reaches a peak. In other words, construction is responsible for the output of buildings and infrastructure upon which most other economic activities depend.

Once an economy has developed, the demand for construction products declines and construction output as a percentage of GDP tapers off. In an industrialised nation the building stock is well developed, so the need to add to it is less. Much of the infrastructure and many of the buildings may be ageing, but the requirement for new build work is generally smaller – however, there is likely to be a far greater need for repair and maintenance work. Generally, therefore, the higher the GDP per capita, the higher the proportion of repair and maintenance work in the construction sector. This relationship is described by the Bon curve (see Reading 5, page 261).

An interesting contrast is offered by comparing construction output in Western Europe and the states that comprised the former Soviet Union. In Western Europe, the construction market increased annually by approximately 1.5 per cent on average in the period 2000 to 2004 compared with an average annual growth in GDP of 2.0 per cent. In short, GDP growth exceeded construction output growth by 0.5 per cent per year over the five-year period. In stark contrast, the newly independent states of the former Soviet Union (comprising Armenia, Azerbaijan, Georgia, Kazakhstan, Kyrgyzstan, Republic of Moldova, Russia, Tajikistan, Turkmenistan, Ukraine and Uzbekistan) all making their transition towards market economies during the same period, experienced an average annual growth in construction of 10.3 per cent compared with an average annual growth in GDP of 7.5 per cent. In other words, construction increased between 2000 and 2004 at a faster rate than the whole economy (United Nations 2005).

The relationship between GDP and construction output is discussed further in the three case studies that conclude this chapter (see pages 232–6). However, we end this section by entering some general caveats concerning the reliability of GDP data. It is difficult to obtain accurate data on the output of the construction sector for two reasons. First, the construction industry comprises a very large number of small geographically dispersed firms that undertake mainly small projects. This makes compiling comprehensive data sets difficult for government agencies monitoring the industry. Second, alongside the official activities 'put through the books' and recorded in company accounts, there is a significant informal economy – unofficial work carried out for cash in hand. (There is also the associated conundrum of DIY.) Some countries have begun to place a value on this informal output. In France, it is estimated to represent as much as 23 per cent of the official construction activity, in Italy it is 17 per cent, in Spain 15 per cent, and in the UK it is thought to be around 10 per cent (DLC 1998: 28). In Eastern Europe, the informal sector in construction is reported to be even larger, with Hungary estimating that 28 per cent of construction work is not recorded in the official statistics and in Poland unofficial activity is around 20 per cent of total construction output (DLC 1998: 10).

The most extreme examples of unreliable data seem to come from countries in transition. For instance, data derived from the Turkmenistan National Institute of Statistics and Forecasting reported real GDP growth to be 19 per cent per year during the period 2000–2003, which would suggest that Turkmenistan had the fastest growing economy in the world at the time. However, these rates diverge considerably from the estimates for Turkmenistan reported by the United Nations secretariat. The UN estimated an annual average real GDP growth for 2000–2003 of only 8.4 per cent – and these figures were used in the above analysis comparing construction output against GDP.

Key Points 13.1

❏ The construction industry can be defined in many ways. Broad definitions suggest that the construction industry typically accounts for 14–16 per cent of total annual economic activity; on the narrower definition, it only accounts for about 6–8 per cent.

❏ The simple circular flow model highlights (a) that households sell factors of production in return for incomes, (b) that businesses sell goods and services to households, and (c) that there is a close relationship between income, expenditure and output.

❏ National accounts measure the annual level of economic activity, and economic growth is identified by changes in 'real' GDP.

❑ GDP represents the total money value of all production created within a country during a year. GNI includes the income generated for the nationals of that country by overseas activities.

❑ There is a relationship between the level of construction activity and a country's stage of development. Construction can be regarded as the engine of economic growth.

❑ In most countries, informal output means that construction activity is under-reported in national accounts.

FROM CIRCULAR FLOW MODE TO REALITY

The two-sector circular flow model presented in Figure 13.1 suggests that the amount of money flowing around an economy is always constant – the GDP figures never change; economic growth is always equal to zero. This is because the model is based on the assumption that expenditure levels precisely determine income levels, and expenditure is in turn determined by income, and so on. In this theoretical model of a two-sector economy, income and expenditure levels are permanently static: there is no growth and no decline. This economy could be classed as being in neutral equilibrium.

In reality, however, every economy experiences leakages (withdrawals) from the circular flow. These occur through the sectors we initially specifically excluded: overseas, financial institutions and government. Simultaneously, there may be injections of funds into the economy through these sectors, for example from exports (that is, earnings flowing in from abroad).

Leakages and Injections

Figure 13.2 (on page 224) extends the circular flow model to include leakages and injections. Three leakages from expenditure are shown: savings, imports and tax. Counterbalancing these leakages, there are three injections: investments, exports and government spending. The decisions that determine the overall size of these leakages and injections of funds are carried out by different groups of individuals with different motivations. It is most unlikely, therefore, that leakages and injections will be equal and cancel one another out.

If the total level of leakages is greater than the level of injections, the economy will become run down, raising unemployment and reducing standards of living. To take an extreme example, if every household decides to spend its money on imports from abroad, then this would represent a major leakage of funds from the domestic circular flow and a significant boost to other countries. Conversely, if the total level of injections is greater than the level of leakages, the economy will be boosted, increasing employment opportunities and raising the amount of national income.

Figure 13.2 The circular flow model with injections and leakages

To complete the circular flow diagram, leakages need to be subtracted as households save, spend money on imports, and pay taxes to the government. And injections need to be added as businesses benefit from investment funds, export earnings and government spending.

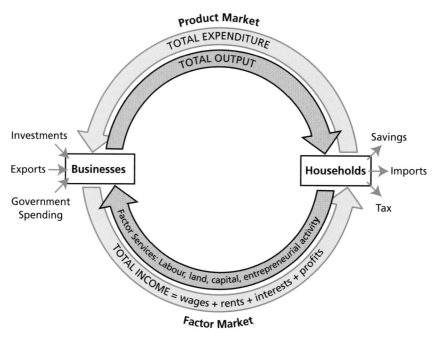

EQUILIBRIUM OF THE MACROECONOMY

Equilibrium means a 'balanced state'. In the macroeconomic context, this means that income, expenditure and output levels continually adjust upwards and downwards to keep in line with one another. For example, when leakages exceed injections, expenditure on domestic output will be less than factor incomes. Consequently, firms will not receive sufficient revenue to cover their output costs. Stocks will accumulate and firms will cut back output and incomes until they equal expenditure again. A new level of equilibrium will have been reached.It is the nature of imbalances between leakages and injections that prompt changes in output from year to year. These changes lead to different levels of income circulating within the economy – representing different levels of economic activity. The dilemma for economists, however, is that although all economies tend towards equilibrium, the associated level of activity is not always sufficient to support full employment.

It should be clear by now that national income analysis is one-dimensional, in so far as it focuses entirely on the monetary value of material goods and services. Whenever we discuss leakages and injections, the environmental dimensions (as covered in Part B) are ignored. The interaction between the economy and the environment is not effectively measured in conventional national accounts. In fact, any money spent on cleaning up the environment from pollution simply contributes to total expenditure in just the same way as money spent on dumping waste, dealing with crime or preparing military attacks. At present, all output is regarded as contributing to economic wellbeing. In historical terms, an explanation for this positive attitude to all output is due to its association with employment. This will become clearer in the next section.

Key Points 13.2

❑ Exports, investments and government spending represent injections to the circular flow. Imports, savings and tax represent leakages. The size of the injections set against the leakages determines the annual level of economic activity.

❑ All economies tend towards equilibrium but that does not guarantee full employment.

❑ Conventional national accounts do not systematically measure the interactions between the economy and the environment.

MANIPULATING THE LEVEL OF ECONOMIC ACTIVITY

In the last 60 years, there has been some debate between economists about the policies that should be adopted by governments to manage the economy. The discussion has proceeded in two main phases. During the years following the Second World War, the consensus of opinion seemed to be for an interventionist strategy. This approach was informed by Keynesian **demand management** theory. The major objective was to keep the economy running near full capacity without incurring wide fluctuations in output. During the last twenty or thirty years, the Keynesian consensus has given way to **supply-side economics**, with unfettered free markets providing the theoretical basis of the approach. The dominant economic objective of supply-side economics is to control inflation in order to achieve high and stable levels of growth and employment. These two contrasting approaches to managing the level of economic activity are considered below.

Demand Management

When a government is faced with a situation in which resources are unemployed and the economy is generally running below full capacity, it can intervene in various ways to reflate the economy. The easiest option is to increase its own spending, and thereby inject funds into the circular flow. This idea, known as demand management, was fashionable throughout Europe from 1945–1975. It was an attractive option because injections of government funds were seen to have a multiplier effect on national income level and employment.

THE MULTIPLIER THEORY

The theory of **multiplier** builds on the circular flow concept – on the idea that expenditure determines the level of output and its associated income. In other words, when people are employed they spend their wages on goods and services produced in other sectors of the economy which, in turn, generate employment and spending elsewhere, creating an upward spiral. Keynes argued that if the current amount of expenditure is insufficient to maintain full employment it becomes advisable for the government to intervene – or, to express it in journalistic terms, to 'kick start' or 'pump prime' the economy.

Consider this scenario: assume a government invests £40 million for a new road. This will cause expenditure and output to raise by the same amount. To increase output, more labour will be taken on. New firms may be started. The newly employed resources will be rewarded with incomes to the value of the initial injection. However, as this money circulates around the economy, some of the £40 million will leak out of the flow in the form of savings, imports or taxes. Economists refer to this as the **marginal propensity to leak** (MPL). The concept of the margin – as we discussed in Chapter 7 – focuses on additional or incremental amounts. The marginal propensity to leak, therefore, represents the proportion of the 'additional' income that does not get used on consumption. If we assume an MPL of 25 per cent, we can quickly calculate that households will spend £30 million of their increased income on **consumer goods**. (Certainly if the recipients of income injected by government spending were previously unemployed, we would expect these households to spend any additional income coming their way on consumption rather than saving.) This additional spending will add a further boost to total expenditure. In turn, firms producing consumer goods will increase output, and they will take on more resources and have to pay out more in interest, wages and rent in order to earn more profit. Again incomes will increase. This will lead to successive rounds of further expenditure. If we continue to calculate the increase in additional expenditure occurring as a result of the initial additional government investment of £40 million, we find that national income is 'pumped up' by a significantly larger amount. In this example, it would actually be £160 million. The determining factor is the size of the leakage, since the multiplier is equal to the reciprocal of the MPL. In developed European economies the leakages are quite large and, accordingly, the multiplier effect is significantly smaller than our example suggests.

This scenario is not too far removed from the immediate post-war reality. The government policy was to invest in the construction industry to increase the general volume of economic activity. Construction was specifically chosen because in most countries it was, and still is, a labour intensive activity, and it plays an important role in the development of the productive capacity of the economy. In fact, many post-war economists regarded the construction industry as a 'regulator' of the economy. The European Commission (2001: 7) has estimated that one job in construction gives rise to two further jobs in the economy as a whole.

Aggregate Demand

A central part of Keynesian analysis is **aggregate demand** (AD).

> Aggregate demand can be defined as the total spending on goods and services produced in a whole economy.

At the beginning of this chapter we considered total expenditures on a theoretical level in a two-sector economy. In such a model, aggregate demand would be equal to consumption expenditures (for example, on beer and chocolate) by households and investment expenditures (for example, on buildings and machinery) by businesses. In reality, however, we need to add government expenditures (such as on road construction) and export revenue from UK output (such as US purchases of Jaguar cars). Aggregate demand (AD), therefore, consists of four elements: consumer spending (C), investment spending (I), government spending (G) and expenditure on exports (X). Aggregate demand is the total of these four elements once one further adjustment is made: to be technically correct, spending on imports (M) from abroad needs to be subtracted as this is not money spent on UK products. It is traditional for the shorthand notation to be used to express aggregate demand using the formula:

$$AD = C + I + G + X - M$$

At this juncture you could be excused the feeling of *déjà vu*. Earlier in this chapter we discussed national income accounting and derived a monetary value for economic activity. Aggregate demand is in fact analogous to GDP. Table 13.2 shows the components of UK aggregate demand in 2006.

Table 13.2 Measuring aggregate demand in 2006 market prices

Components of aggregate demand	C + I + G + X − M = AD
Related amount in £bn during 2001	823 + 239 + 287 + 370 − 424 = £1300 billion

Source: United Kingdom National Accounts (ONS 2007)
Note: All figures rounded to nearest £billion

Demand management techniques proved to be a difficult tool to use. One of the difficulties was the timing of the action. It becomes particularly difficult when the sector used for delivery is construction, as the time lags tend to be long and variable. A second and more obvious problem was overshooting – adding a too large injection which causes the economy to overheat. The subsequent excess demand achieves nothing except continually increasing prices – resulting in higher inflation. In short, it proved very difficult to use demand management techniques to shift aggregate demand to the precise level to secure full employment at the right time.

These problems were neatly analysed by Professor Frank Paish during the 1960s. He suggested that the problem with Keynesian demand management techniques lay with the concepts of productive capacity and actual output. He argued that by adding to productive capacity, you generate income before you generate actual output. Therefore, to achieve stability, Paish recommended maintaining a margin of unused productive capacity.

The focus has now shifted to concerns about potential output and actual output. Once spare capacity has been used up – and full employment of resources has been achieved – actual output will be restricted in the short-term. The answer to the problem, therefore, seems to involve ensuring an amount of unused productive capacity and/or increasing the potential output of an economy. To increase the level of output, at the full employment point, requires more capacity and this is determined by resources being used efficiently. Technically speaking, economists had become concerned about the supply side. Patricia Hillebrandt (2000: 66) has always regarded the capacity of the construction sector as a constraint and she has identified three episodes – in 1964, 1973 and 1989 – when the industry or its material suppliers were stretched to capacity. Each episode resulted in periods of sharp inflation, and the latter two also led to periods of recession in 1974 and 1991.

Key Points 13.3

❑ During the years immediately following the Second World War, the strategy for manipulating economic activity was informed by Keynesian demand management theory.

❑ Governments wishing to slow down or speed up the rate of economic activity used demand management policy to affect the size of injections or leakages.

❑ An increase in government expenditure causes a multiplier effect on the level of national income. The larger the marginal propensity to leak, the smaller the multiplier effect.

❑ Aggregate demand is the sum of all expenditures in an economy. It is usually considered in four categories, and using the standard notation, defined by the formula C + I + G + X – M.

❑ Demand management techniques are closely associated with the problem of inflation.

SUPPLY-SIDE ECONOMICS

One of the main hallmarks of economic policy since 1979 has been concern over the supply side of the economy. The focus has shifted from government spending and aggregate demand to production and aggregate supply. **Aggregate supply** can be regarded as total production, and clearly many factors influence its size such as the level of profits, ease of movement into and out of markets, the level of wages, the efficiency of capital and labour, the level of fixed costs, etc. As a result, policy oriented to the supply side has given rise to measures to increase incentives within the economy. Indeed, economists concerned with this perspective seem to place more trust in market forces than government intervention. The *Statement of Intent on Environmental Taxation* (HM Treasury 1997) was an interesting example of this approach. This set out the UK government's intention to use the tax system to promote sustainable growth, by shifting the tax burden away from 'goods', such as employment, towards 'bads', such as pollution. The policy sought to achieve a 'double-dividend' – of increasing capacity and reducing environmental damage.

Supply-side policy, therefore, is geared to making markets work more efficiently. This has been achieved by reducing the structural rigidities that clutter many markets. For example, **wages councils** were stripped of many of their powers in the 1980s and formally abolished in 1993, trade union activities have been restricted and market competition has been opened up through privatisation and deregulation. Similarly, governments have made broad reductions in rates of income tax – from 33 per cent to 20 per cent at the basic rate and from 83 per cent to 40 per cent for the top rate – to provide more incentives for people to work harder. Indeed, there seems to be no market that has escaped from the plethora of supply-side measures.

Aggregate Supply

By devising supply-side methods to promote competition in as many markets as possible, governments believe they can reduce the money they spend on direct

Figure 13.3 The aggregate supply curve

At a price level of P_1, the AS curve is a horizontal line up to the point where output is Y_1. Then, there is an intermediate stage – some sectors of the economy are experiencing excess capacity, but others are not. At Y_2 there is no immediate capacity in any sector of the economy and prices rise. In the long run, however, it should be possible to increase productive capacity and shift the AS curve towards the right.

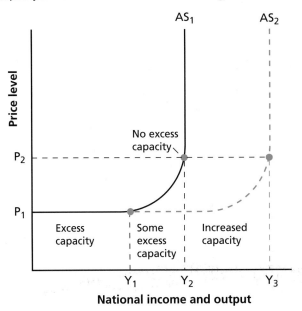

National income and output

intervention and encourage the entrepreneurial spirit that drives production. In technical terms, this policy is designed to shift the aggregate supply curve to the right. This is demonstrated in Figure 13.3. An aggregate supply curve represents the relationship between the output firms would be willing to supply and the general price level.

It would be nice to simply conclude that the aggregate supply curve slopes up because, the higher the price level, the greater the incentive for producers to produce more. But it must be emphasised that we are not talking about changes to individual specific prices – the vertical axis represents changes to the general price level; it is an index of the weighted average of all prices. In order to understand the true purpose of the aggregate supply curve, we examine four situations:

1 large amounts of unused capacity
2 full capacity
3 an intermediate range between the two
4 increasing capacity in the long run.

UNUSED CAPACITY

When an economy has many factories operating at less than capacity, there is a general under-utilisation of the productive capabilities and it is possible to increase output without any pressure on prices. If there is unused capacity, producers can increase supply without having to pay higher prices for factors of production. If they need more labour, they can hire someone who is unemployed – they do not need to pay higher wages to attract people. Providing there are significant levels of unemployment and unused capacity, per-unit costs of output remain the same, no matter what the volume of output. In these circumstances, we would expect the aggregate supply curve to be a horizontal line at the current price level. Consider a current price level of P_1, as given on the vertical axis of Figure 13.3. The horizontal line labelled 'excess capacity' represents that part of the aggregate supply curve AS_1 where an increase in output causes no pressure on prices. Within this range, supply is perfectly elastic.

NO EXCESS CAPACITY

Now consider the other extreme, in which the economy is running at full capacity and there is no unemployment. In such a situation it is impossible for any additional output to be produced. There is only one thing that can happen – the price level can rise, but no further increases in output are physically possible. The aggregate supply curve has to be a vertical line, as shown at output rates Y_2 and Y_3 in Figure 13.3. At this point supply can be said to be perfectly inelastic; any price rises above level P_2 produce absolutely no changes to the quantity supplied.

INTERMEDIATE RANGE

When there is some excess capacity in parts of the economy but no excess capacity in other parts of the economy then as production increases so-called bottlenecks or **supply constraints** may develop. As firms try to increase output they may experience shortages of some inputs, most frequently in certain kinds of skilled labour. When this happens, firms can try to attract more of the scarce input by offering a higher price. They compete with each other for the limited supply of people with scarce skills, thus driving wage rates up. This raises their costs of production, and they then react by raising their prices whenever possible.

The shape of the aggregate supply curve in the intermediate range is explained by these bottlenecks. As the aggregate supply curve starts to slope up, it becomes steeper as full capacity output is approached because more and more supply constraints appear. These constraints force some prices to increase and, as output nears full capacity, sellers can put prices up without losing customers. Since the price level is an index of all prices, if some prices stay constant and some go up, the general price level will rise too. This means that if we increase production from output Y_1 in Figure 13.3 – at the end of the excess capacity output rate – price levels will rise along with national income. In this range, there is a positive relationship between national income and price level. As supply constraints become greater, supply becomes less and less elastic. Successive increases in spending lead to smaller and smaller increases in output and income, up until we reach full capacity.

INCREASING CAPACITY IN THE LONG RUN

In the long run, as technology advances and the stock of capital increases, it will become possible to increase capacity and produce more – so the vertical line showing full capacity output will shift gradually to the right. This is shown by AS_2 in Figure 13.3. The increases in national income and output from Y_2 to Y_3 are achieved by new capacity. This represents how, over a fairly long period, output can begin to increase and prices fall.

As far as construction is concerned, manipulating the supply side has led to the government's role becoming less transparent. Today, governments increasingly rely on subtle messages to improve the market efficiency of the construction industry by promoting ideas such as lean production, factory-built modular structures and partnering. The aim is to encourage construction to embrace technology and vertical integration, and to capture some of the advantages achieved by manufacturing at the site level. If the capacity of the construction industry could effectively increase, the overall aggregate supply curve would certainly shift to the right, enabling total output to increase without putting pressure on resources. This is because construction influences the capacity of so many other sectors of economic activity. The best evidence of increasing capacity and growth through hi-tech construction methods can be seen in Japan, which is why it is one of our three national case studies that conclude this chapter.

Obviously, governments want the aggregate supply curve continually shifting to the right to achieve sustainable growth and stable prices. Unfortunately, however, this is frequently complicated by a construction industry that seems wedded to low-tech approaches.

Key Points 13.4

❑ Since 1979, economic policy has been concerned with the supply side of the economy. The focus of government activity has shifted to production and aggregate supply.

❑ Economists have become increasingly concerned with the principles of microeconomics that underpin aggregate supply.

❑ Supply-side economics revolves around the freeing up of markets and the reduction of direct intervention by the government.

❑ The aggregate supply curve describes several stages based on different levels of capacity. The nearer the economy gets towards full capacity, the more likely inflation becomes (see Figure 13.3).

THREE CASE STUDIES

There is a close relationship between a county's construction sector and its level of GDP. This is partly because the construction industry involves the assembly of many different products from a large number of industrial sectors. As we suggested in the introduction to this chapter, construction activities have far broader impacts than the main official statistics imply. Of greater importance, however, is the fact that construction provides the investment that underpins development, as it provides houses to live in, buildings to work in and infrastructure to support communication and transport. In fact, several studies have shown that the construction industry has significant linkages with other sectors of the economy. Consequently, it has become common practice to use the construction industry as some kind of regulator for the overall economy. The following three case studies demonstrate how this may be achieved in different cultures.

CASE STUDY 1: CHINA

The People's Republic of China has been experiencing a building boom for the last 30 years, with overall economic activity increasing by an average of 9.5 per cent annually. Beyond doubt, China has consistently been the fastest growing economy in the world for several years, although the International Monetary Fund forecasts that the country's economic growth will marginally slow down in the next few years. As a result, the Chinese construction industry has faced, and continues to face, a significant increase in demand. During 2007 it used 40 per cent of the world's output of cement and accounted for 90 per cent of the growth in steel demand. Obviously not all the steel went into construction activity, but the figures highlight the scale of the expansion in the Chinese economy. The proportion of value added by construction to GDP increased from 6 per cent to 10 per cent. Employment in the industry more than doubled from a workforce of 17.3 million in 1985 to 40 million in 2007 – to put this in context, note that China has a total population of over

1.31 billion people and a 775 million strong labour force (Han and Ofori 2001; Crosthwaite and Connaughton 2008).

The construction industry is an important element in the transformation of China towards a more market-oriented system. For example, at the Chinese Communist Party meeting in 1994, construction was explicitly identified as one of the four pillar industries that could help China to begin a new cycle of economic growth. (The other three pillar industries were automobile manufacturing, oil and chemical refinery, and mechanics and electronics.) The focus on construction was supported by the establishment of a new government department. The ministry of construction has taken a lead role in implementing new strategies for the industry. These strategies include opening up construction markets, establishing a competitive bidding system, allowing autonomy in state-owned enterprises and eliminating bribery as a means of obtaining contracts, loans and materials. These policies have resulted in a substantial expansion in the number of Chinese construction firms, and they have been regarded as an important motor in driving economic growth. The Chinese government has been keen to avoid foreign domination, and although foreign investment is allowed, it is not permitted in contracts for the supply of water, gas or subway systems.

The Chinese construction industry has clearly made a significant contribution to the development of the Chinese economy, but academics recognise that the process of rapid urbanisation and modernisation means that it still has a long way to go to reach world-class standards. Essential reforms are required in areas such as private investment, use of the price mechanism, quality of materials and equipment, and, most significantly, reducing resource consumption and severe environmental impacts (Chen 1998; Yang and Kohler 2008). Interestingly, however, one of the current projects involves developing the world's largest eco-city at Dongtan. An 86 square kilometre project on an island off the coast of Shanghai, Dongtan is expected to house a population of 80,000 by 2020 and several hundred thousands in the longer term. The buildings will be powered by renewable energy sources and natural sunlight will be designed into public and private spaces. In addition, waste will be recycled and composted. This development was cited in the Stern Review (2007: 437) as a potential world-class example of sustainable urban development.

CASE STUDY 2: SINGAPORE

In Singapore the construction industry accounts for an average of 7 per cent of the economy and employs about 10 per cent of the workforce. Compared to China, the Singapore construction industry is heavily reliant on imported materials and foreign workers. According to Chan (2002: 523), 25 per cent of Singapore's construction output is made up of imported goods and services and 69 per cent of the labour force (that is, about 80,000 people) are overseas workers. This heavy reliance by the construction sector on imports, of one sort or another, places a strain on Singapore's balance of payments. (For a complete understanding of macroeconomic impact of the construction industry, many issues need to be considered including – as Singapore fully demonstrates – the international dimension.) Construction ranks fifth in terms of its contribution to Singapore's GDP. Empirical results, however, demonstrate that following finance and business, the second biggest influence on the country's balance of payments is caused by changes in construction output (Chan 2002: 532). This scale of leakage of resources is usually more common in less developed countries, where deficiencies in local construction capacity and the subsequent dependence on imported materials, machinery and skilled labour notoriously run up foreign debts and cause balance of payment problems. Singapore, however, is one of the most advanced economies in the world; it has certainly progressed beyond the industrial stage towards a knowledge-based economy.

This is the context for the current development of the Singaporean construction industry. And, setting an agenda for the twenty-first century, a government report has identified five important obstacles that the construction industry in Singapore needs to overcome. The report makes recommendations to address in turn each of these obstacles (Ofori 2002: 406).

- The image of the industry should be transformed from being perceived as 'dirty, demanding and dangerous' to one where it is viewed as 'professional, productive and progressive'.
- Construction must change from a 'labour intensive and low-skilled industry' to a 'knowledge industry'.
- Instead of being an industry based on '*in-situ* projects', construction should become a 'distributed manufacturing process'.
- Construction should shift from being an industry based on 'segregated activities' to a far more 'integrated process'.
- Construction should move from being low cost on the basis of low wages to attaining low cost through high production.

These five obstacles could equally apply to the construction industry of most developed economies with, arguably, the exceptions of the United States and Japan. In these two countries, information technology has been used to raise the level of integration, competition, professionalism and productivity. In fact, it has been suggested that in Japan productivity rates in construction are double the rate of efficiency in Singapore (Ofori 2002: 409).

CASE STUDY 3: JAPAN

During the last couple of generations, the Japanese have rapidly developed their country into one of the richest nations in the world and much of this success is attributed to Japan's active construction sector. Indeed, Japan is worthy of a case study not only because the Japanese construction industry is extremely efficient in terms of output, but also because the sector has proved to be 'unusually profitable' (Fraser and Zarkada-Fraser 2001: 831). Furthermore in terms of annual spending Japan currently represents the second largest construction market in the world (Crosthwaite and Connaughton 2008).

In the fastest growing, fully industrialised economies, construction rarely accounts for more than 10 per cent of GDP. However, when Japan was at the peak of its rapid urban development during the 1990s, construction – in the narrowest sense – accounted for about 12 per cent of GDP. The Japanese construction industry is in fact characterised by many exceptional features. Here we just give three of the most significant examples.

- The majority of commercial buildings are constructed for owner-occupiers rather than for speculative developers.
- The relationship between general contractors, specialist subcontractors, labourers and clients is frequently characterised in terms of collaboration and integration rather than conflict and fragmentation. In an analysis of the Japanese construction sector, Reeves (2002: 421) even goes as far as suggesting that the whole mechanism in the sector 'operates to provide mutual benefit to all of the players involved'.
- Firms compete on the basis of technology, in contrast to the usual price-based competition that drives the construction process.

It is this third point that makes Japan's construction industry unique. In most countries, construction is regarded as traditional, conservative, labour intensive and not particularly interested in innovation; in Japan, it is quite the opposite. The Japanese government regards 'hi-tech intelligent' buildings as central both to the nation's infrastructure and to the development of a knowledge-based economy. It therefore promotes technology use in the management of a building.

Technology is also embraced at the site level, where the construction of high-rise buildings utilises prefabrication, robotics, automation and information technology. Indeed, 'computer integrated construction' is becoming a key part of the corporate policy of several major Japanese contractors. As Professor David Gann (2000: 113) explains in his analysis of innovation and change in the global construction market: 'Rich Japanese companies invested heavily in electronic technologies for their new buildings. Moreover, government played a bigger role than elsewhere in sponsoring the concept of intelligent buildings. Japanese construction firms vied with electronics, telecommunications and office equipment manufacturers to produce prestigious buildings.'

This level of technological innovation is associated with a higher than usual commitment to research and development (R&D). In most OECD countries, R&D in construction rarely exceeds 0.4 per cent of the industry's total annual output. In the UK, it rarely exceeds 0.1 per cent. In Japan, it is usually above 1 per cent – and has exceeded 2 per cent. Compared to manufacturing, this is still very low – as generally manufacturing firms invest 3–4 per cent of their total annual output in R&D. The important point for our purposes, however, is that Japan again stands out as the exception to the general rule. It has developed five, very large, vertically integrated construction companies (Kajima, Obayashi, Shimizu, Taisei and Takenaka) that believe in using technology to obtain competitive advantage. Again in contrast with other countries, construction research is concentrated in the private sector and there are more than 20 companies that possess their own technological research institutes. To finish with a historic example, in one financial year (1992–93) Kajima – one of Japan's 'big five' – invested over £125 million in R&D. In contrast, in the same period, all the UK private contractors only managed a combined R&D spend of £17.5 million (Gann 2000: 198).

Nearly 10 years later, the *Rethinking Construction Innovation and Research* report regarded the total UK private sector construction R&D annual expenditure of £40 million as still being far too low given the size and significance of the sector (Fairclough 2002: 13). Interestingly, in the context of this text, Fairclough (2002: 30) emphasised that construction research was inadequate when compared to the important contributions the sector made to the UK's economic, social and environmental wellbeing, and its main conclusion and recommendation was the need for the government to increase its commitment to construction R&D. Fairclough (2002: 30) wanted to create a 'virtuous circle...beginning with more and better focused R&D investment, allowing more innovation, leading to better profitability, and providing the additional capacity to invest in more R&D'. Japanese construction is the exception to the rule as it already has introduced this strategy and broken into the virtuous circle. However, to achieve the same level of progress in the UK, or elsewhere, some government intervention and funding would be necessary.

A role for government

Governments have a central role to play in promoting change in the construction industry. In the period dominated by Keynesian policy (1950–1975) governments opted for direct intervention. Indeed, until the 1970s, 50 per cent of all construction work was purchased by the public sector in the UK. More recently, however, government policy has placed far greater reliance on laissez-faire market forces. This has eroded the capacity of governments to directly control construction output. Consequently, most governments today facilitate changes within construction by acting as a catalyst or by adopting co-ordination functions. As we have argued, however, the fragmented nature of the industry is generally a barrier to change – and this raises significant challenges. It certainly makes it more difficult to control output through these indirect methods and it is a constraint on the introduction of more energy efficient and environmentally friendly practices.

There are also problems posed by economic growth. The rapid economic growth rates in China and India mean that Asia is destined to become responsible for more global CO_2 emissions than any other continent by 2010, driven largely by China and its heavy use of coal for power generation (Stern 2007: 201). The proportion of economic growth in these two developing countries, also means that Asia is forecast to account for more than half of the global market for construction services by 2030 (Chevin 2008: 1).

Modern governments have attempted to exercise their responsibility to the industry through a variety of policies, including encouraging environmental protection, promoting competition, innovation and R&D, setting regulations relating to health, safety and building standards, and being a catalyst for promoting change. Initiatives to encourage sustainable construction are an interesting example of how modern governments have attempted to manipulate activity and these are reviewed in the concluding chapter of this book.

> ## *Key Points 13.5*
>
> ❑ Regardless of the level of economic development, and a nation's culture and tradition, the construction sector plays an important role in any economy.
>
> ❑ In China, Singapore and Japan, the construction industry has provided an engine for economic growth.
>
> ❑ Japan's construction investment in R&D is the highest of all OECD countries. In most countries, however, it is left to governments to initiate policies to promote long-term goals.

MEASURING ENVIRONMENTAL IMPACTS

As indicated in Chapter 12, for the past 50 years or more academics have debated how to establish a standard international measure that effectively monitors the broad usage of environmental resources. This is now becoming increasingly urgent. Various ad hoc measures have been trialled, and many of these defer to GDP as a reference point. For example, by expressing changes in real GDP in relation to energy used (measured in million of tons of oil equivalent), it is possible to calculate the amount of energy required to support a certain level of GDP or, more simply, the energy consumed per unit of output. With efficiency gains it is possible for energy, and emissions, per unit of output to decline. The formal way to refer to this phenomenon is to say it is possible to 'decouple' growth from energy use, or in plain English to accept that it is not inevitable that economic growth always produces an equivalent deterioration in environmental quality.

Research into resource usage has been progressing since the 1960s, and the final part of the *UK National Accounts* is now dedicated to environmental issues by detailing oil and gas reserves, atmospheric emissions, energy use, etc. Much of the data in the section is detailed in units of physical measurement or volume and this leads to debates about its use or value. While there is some agreement that governments need a yardstick to monitor and measure the total environmental impacts placed by humanity on to the ecosystem, there is no accepted consensus about the best way forward. In short, we need an authorised metric designed to capture environmental activity, which functions in a similar way to GDP in capturing economic activity.

The metric that currently perhaps has the best potential to meet this role is the **ecological footprint**. Accounts based on this metric already exist for 150 countries for the years 1961 to 2003. They are produced by an organisation called the Global Footprint Network, which uses data provided by the United Nations. A major source of interest and funding for this work comes from the EU. The Global Footprint Network's vision is to make the ecological footprint as prominent a metric as GDP by 2015.

The basis of the system is to provide a way of measuring how much productive land is required to sustain life in terms of producing food, energy, water and timber and carrying away the associated wastes. The calculation of the demands we place upon the planet can be made at the individual, national and global levels. This is done by taking resource use and comparing it to what is actually available. The convention is to state the data with reference to the global hectares available per person. This is explained in the next section.

Measuring the Ecological Footprint

The calculation begins with the assumption that productive land upon which life depends could be anywhere in the world. This seems reasonable, as people of one nation usually consume resources and ecological services from all over the world. The ecological footprint calculations are, therefore, stated in global hectares (gha). A global hectare is one hectare of biologically productive space. In 2003, the globe had 11.2 billion hectares of biologically productive area corresponding to roughly one-quarter of the planet's surface. These 11.2 billion hectares include 2.4 billion hectares of water and 8.8 billion hectares of land. The land area is composed of 1.5 billion hectares of cropland, 3.4 billion hectares of grazing land, 3.7 billion hectares of forest land, and 0.2 billion hectares of built-up land. Non-productive marginal areas, such as deserts, ice caps, and deep oceans, are not included in the 11.2 billion global hectares of biologically productive space (GFN 2006).

To express the capacity available to each person, the 11.2 billion global hectares of biologically productive area are divided by the number of people on earth. The world's population was approximately 6.3 billion in 2003, giving 1.8 global hectares per capita. This means that, in principle, the average amount of ecological productive capacity (the biocapacity) that exists on the planet per person is 1.8 global hectares. (This simple calculation of available biocapacity assumes that no capacity is set aside for the demands of wild species.)

If each person could survive on 1.8 hectares, or less, the world would be sustainable. However, the latest ecological footprint accounts published by the Global Footprint Network indicate that we are far away from achieving sustainable development. The figures show that, on average, every person on earth is taking 25 per cent more from the planet than it can naturally regenerate. In other words, humanity's ecological footprint is 25 per cent larger than the planet's capacity to produce these resources. This means that there is an ecological overshoot – it now takes about one year and three months for the planet to regenerate what we use in a single year. In the language of sustainable development, to meet today's needs we are raiding our children's inheritance, leaving them with less for the future. If all human beings adopted lifestyles of the high-income countries of Western Europe and the US, the world would be living way beyond its means. To sustain the entire world's population with a lifestyle currently enjoyed by the average US citizen would require a productive area equivalent to five planets.

A brief summary of ecological footprint data is presented in Table 13.3; for comparative purposes, this table shows results for the same countries as those selected in Table 13.1 (on page 220). Analysing the two tables, therefore, indicates the resources that are being consumed to maintain the present levels of economic activity (GNI). The figures must be read with considerable care. It is evident that on a *per capita basis* the low and middle-income countries, represented here by India and China, have smaller ecological footprints than high-income countries. However, because of their extremely large populations, both India and China have a far bigger *total* ecological footprint than many high-income countries, such as Japan, Norway and the UK. Table 13.3 ranks the countries according to their total ecological footprint. In fact, two countries alone – the United States and China – are responsible for over a third of the world's ecological footprint; 35 per cent on the basis of 2003 figures and possibly higher today.

Table 13.3 The ecological footprint of selected economies

2003 Data	Population	Total ecological footprint	Per capita ecological footprint	Per capita biocapacity	Ecological surplus (deficit)
	(millions)	(million gha)	(gha/person)	(gha/person)	(gha/person)
World	6,301.5	14,073	2.2	1.8	-0.5
United States	294.0	2,819	9.6	4.7	-4.8
China	1,311.7	2,152	1.6	0.8	-0.9
India	1,065.5	802	0.8	0.4	-0.4
Japan	127.7	556	4.4	0.7	-3.6
United Kingdom	59.5	333	5.6	1.6	-4.0
Norway	4.5	26	5.8	6.8	0.9

Source: Adapted from National Ecological Footprint Accounts, 2006 edition.

Another way of assessing sustainability at a national level is to compare a country's consumption against its national biocapacity – that is, the resources it can produce sustainably (including managing wastes) within its own borders. A national biocapacity can be expressed in terms of global hectares, or global hectares per person in the country being considered, and national per capita biocapacities for our featured six countries are shown in the fifth column of Table 13.3. Apart from Norway, all the countries in Table 13.3 exceed their own biocapacity. As the final column shows, most countries have a negative ecological surplus – in other words, they have an ecological deficit. This ecological deficit measures the amount by which a country's footprint exceeds its biocapacity. Only about a third of the 150 countries for which a footprint has been calculated show a ecological surplus – they have some ecological reserve as not all their national (biological) capacity is used for consumption, or absorption of waste, by their own populations. This ecological reserve, however, does not necessarily remain untouched, as it may be used by other countries through production for export.

Most countries' biocapacity (determined by the resources available within their borders) is smaller than their ecological footprint (which is based on the resources that they actually consume). You might think that this is particularly the case in high-income countries, and certainly the United States and most nations in Western Europe have significant ecological deficits. However, some developed countries, such as Norway (as Table 13.3 shows), Canada and Australia, have ecological surpluses because they have relatively abundant resources. Conversely, many countries in the developing world have ecological deficits, because they are resource poor or because they have large populations relative to their resource base.

This paints rather a daunting picture, as the only way to reduce a nation's ecological footprint, or more importantly the global ecological footprint, is if populations become smaller, or if we can find a way to reduce the average consumption per person and increase resource efficiency through changes in technology. The latter could be achieved in part by using fewer fossil fuels to reduce carbon emissions and becoming more reliant on energy from renewable sources.

THE ECOLOGICAL FOOTPRINT AND CONSTRUCTION

The scale of the demands made on the natural environment is easily exemplified by construction, as the sector (on the broad definition) consumes resources and generates waste on a scale that completely dwarfs other sectors of the economy. To begin with, it is the environment that provides the land on which buildings and infrastructure are located. Subsequently, it is the environment that supplies many of the resources that are used to make building material products. Finally, it is also the environment that is ultimately responsible for assimilating and processing the waste that arises from the various phases of a property's life, from construction through to demolition. The scale of the problem was discussed in Chapter 11 (see pages 181–2). There we considered the resource intensive nature of construction inputs and outputs, and discussed the problems of an industry that produces 3 tonnes of construction waste for each man woman and child in the UK.

Changes in the ways that property is designed and used could have a significant impact on the scale of the ecological footprint. At present, a large part of the ecological footprint is generated through energy usage, as carbon emissions make up about half of the total footprint. These could be reduced to zero by designing buildings to harness renewable sources, so-called carbon-neutral solutions that rely on wind, sun or geothermal energy sources. Similarly, the **embodied energy** that accounts for a significant amount of the energy used in the construction phase could be reduced by recycling materials, conserving buildings to reduce the need for new build, and sourcing materials locally to avoid unnecessary transportation.

Three Scenarios

Scenarios are often used by economists to explore logical consequences of different sets of conditions or choices. The World Wide Fund for Nature (WWF) has developed a tradition of reviewing ecological footprint accounts in terms of three scenarios: business as usual, a slow shift and a rapid shift. The business-as-usual scenario is based on steady growth of economies and population and it suggests that by 2050 humanity's demand on nature would be twice the biosphere's productive capacity. At this level, the collapse of the ecosystem is highly likely. The other two scenarios however suggest the possibility of achieving sustainability, either as a slow shift from the current development path or a more rapid transition towards sustainability.

The slow-shift scenario suggests that humanity's ecological footprint could be reduced by 15 per cent from 2.2 global hectares in 2003 to 1.5 global hectares in 2100, providing that 50 per cent cuts can be achieved in global CO_2 emissions, etc. The rapid-shift scenario, however, could result in an elimination of overshoot by 2080 and a modest 10 per cent biocapacity buffer which could be used by wild species. This scenario assumes that the demands made on the ecological footprint could be reduced by 40 per cent from 2.2 global hectares in 2003 to 1.3 global hectares in 2100. Achieving this would require significant reductions in CO_2 emissions, with cuts of 50 per cent by 2050 and 70 per cent by 2100 and improved technology to achieve a 30 per cent growth in the capacity of cropland, fisheries and forest yields (WWF 2006).

The three WWF scenarios are distinct from previous economic exercises due to the focus on ecosystem services and their ability to support humanity. The scenarios begin in 1960 and run until 2100. Figure 13.4 suggests the outcome of the alternative approaches.

Figure 13.4 Three ecological footprint scenarios, 1960 to 2100

Source: WWF, 2006: 3

A similar line of analysis was presented by Nicholas Stern (2007: xv) in his international report on *The Economics of Climate Change*. He concluded that the cost of reducing carbon emissions was likely to be about 1 per cent of global gross domestic product, while inaction would lead to a cost of between 5 and 20 per cent of global GDP. In short, the costs of a business-as-usual scenario are extortionate and far outweigh the savings that could be achieved by immediate government action.

CONCLUSION

The ecological footprint metric is still being developed and refined. Adopting this measure – or something similar – will be controversial. It forces governments (and people) to compare the current ecological demands placed on the planet with the capacity of the earth's life-supporting ecosystems. It strongly suggests that we are not living within the sustainable limits of the planet. Ecosystems are suffering, and the further we continue down this path of unsustainable consumption, the more difficult it will become to protect (and restore) the planet's biodiversity. Governments will be unable to address this question properly until they have an agreed way to monitor and measure the problem. The challenges are huge. They are addressable, but it will require concerted action at local and international levels and significant changes in resource intensive sectors such as construction.

Key Points 13.6

- ❑ The ecological footprint measures the natural resource consumption required to support the existing standard of living. Calculations have been carried out for 150 nations.

- ❑ The Global Footprint Network's vision is to make the ecological footprint as prominent a metric as the gross domestic product (GDP) by 2015.

- ❑ The global ecological footprint changes with population size, average consumption per person, resource efficiency, changes to technology, recycling, and attitudes to conservation.

14 The Business Case: Inflation and Expectations

In any government literature that aims to encourage sustainable construction, the key argument is the business case. This makes sense, as the private firms – the primary target for these government appeals – are first and foremost in business. The bottom line means that companies have to earn enough to pay their bills. Consequently, although the ultimate aim of the policy on sustainability is to create a more socially and environmentally responsible construction industry, the business case is seen as the way to drive this forward. The UK government *Strategy for Sustainable Construction* made the point explicitly in its opening pages when it stated that increasing profitability by using resources efficiently lies at the heart of the case for sustainable construction (HM Government 2008: 5). This requires a stable and competitive economy, run in line with the macroeconomic objectives set out in Chapter 12 (see Key Points 12.1).

At the heart of a stable and competitive economy lies a low inflation rate, and the explicit aim of monetary policy since 1997 has been the control of inflation. If prices are continually changing, entrepreneurs are hesitant to enter into contracts as they cannot work out the long-run results of their investments. This is compounded by the problems caused by changing interest rates and fluctuating foreign exchange rates, which often accompany inflation. It is simply much easier to work within a stable economic environment. Stability means that the costs and prices of any project or investment can be estimated with greater certainty and transparency, allowing businesses to plan with increasing confidence. Indeed, economists define the associated impacts of inflation on business as **menu costs**. These costs arise due to the need to revise existing contracts, new tenders and bids for work as inflation sets in. Obviously, as inflation rates become higher and more volatile, menu costs get more demanding and, in extreme circumstances, they might include costs associated with alterations to vending machines, the costs of printing revised price lists, the time spent renegotiating, and so on.

Inflation also affects those that are not economically active. If prices increase, then those on fixed incomes such as pensioners, students and people on state benefits suffer. In consequence, many aspects of economic activity are index-linked to allow for inflation. For example, savings, business contracts and pensions can all be adjusted in the light of inflation. All that is needed is a reliable price index.

INFLATION AND HOW IT IS MEASURED

Inflation is a *persistent* increase in the general price level.

The italicised word in the definition is important, as any increase in the price level must be sustained to be categorised as an inflationary situation. Continuous annual price rises, however, such as those experienced in the UK, are definitely categorised as **inflation**.

Table 14.1 shows UK annual average inflation rates for the period 1950–2005. It highlights that the 1950s and 1960s were periods of low inflation, with prices rising by less than 5 per cent in most years. In the 1970s inflation rates were much higher and very volatile – in fact, during 1975 alone prices rose by 27 per cent. The 1980s saw the second highest rises in consumer prices in the post-war period, and much of this reflected the increased costs in mortgage payments and council tax. The most recent period shown in the table, from 1990 to 2005, could be characterised relative to the previous two decades as a period of **low inflation** with rates rarely exceeding 2.8 per cent.

Table 14.1 UK inflation rates

Period	Annual average change in price
1950 – 1959	4.0%
1960 – 1969	4.0%
1970 – 1979	13.7%
1980 – 1989	6.6%
1990 – 1999	3.7%
2000 – 2005	2.5%

Source: Office for National Statistics

Deflation, which is defined as a sustained fall in the general price level, is the complete reverse of inflation, even to the extent that it typically results from a fall in aggregate demand, in response to which firms reduce prices in order to sell their products. Inevitably, as a form of price instability, deflation also has adverse economic consequences and its poses particular difficulties for economic management – one being that as interest rates cannot fall below zero, the Bank of England monetary policy committee would not be able to reduce interest rates to stimulate demand. The macroeconomic aim, therefore, is to steer a stable course between inflation and deflation. To achieve price stability, the current target is to maintain the UK inflation rate around 2 per cent – and it certainly should not be allowed to go below zero.

Measures of inflation involve representing changes in price over a period of time. The statistical device best suited for this purpose is index numbers. Index numbers are a means of expressing data relative to a given **base year**. They enable the cost of a particular range of products to be expressed as a percentage of the cost of the same group of products in a given base year. The basic principle is shown in Figure 14.1. What is put into the basket for the base year and comparative year must be the same. That is, the system is dependent on like-for-like price comparisons, or as near as possible, from one time period to the next.

This approach is used to produce construction industry indices. For example, the Investment Property Databank compiles a property price index by comparing specific capital and rental values for 11,400 commercial properties on a regular

Figure 14.1 Calculating a price index

In this example there are two identical baskets of goods. The base year is 1997. The price index of 2003 compared to 1997 ends up as 117. In other words, the basket of goods has increased in price by 17 per cent during the period 1997–2003.

Price Index $= \dfrac{2003}{1997} \times 100$

Price Index $= \dfrac{£35.60}{£30.48} \times 100$

Price Index $= \qquad 117$

basis. This data goes back to a base year in 1971. Similarly, the largest two mortgage lenders – Nationwide and Halifax – publish a house price index each month, based on comparisons of the prices of four different house types across 13 regions of the UK. These two financial institutions base their indices on the mortgage offers they have made in the previous month. Houses costing over one million pounds are excluded, as it is assumed that this may distort the average picture. A more comprehensive but less up-to-date measure comes from the Land Registry. This index is based on stamp duty transactions and, therefore, covers every property deal in England and Wales. However, it is slow to reveal what is happening in the market since the stamp duty is paid at the end of the property transaction and that can be four or five months after the sale price has been agreed – and house prices can rise or fall by as much as 20 per cent within three months. Recently, two websites – Hometrack and Rightmove – have added further indices to this set. They collect information from 4,000 estate agents on asking prices and agreed prices and cover more properties than Nationwide and Halifax. All these indices are monitored and collated by the Department for Communities and Local Government (DCLG) to form a further measure of house prices.

Each house price index produces a different estimate of house price inflation. This is because the indices are dependent both on the sample data that has been used to make up each specific basket and the way that each index is subsequently calculated. For interest, we contrast the DCLG data with the Nationwide and Halifax indices in Figure 14.6 (see page 258).

House price indices are important because they are used by the government and the monetary policy committee as an indicator to assess the economy. For a discussion comparing their strengths and weaknesses, see the Office for National Statistics methodological commentary by Dey-Chowdhury (2007).

Many other institutions and professional bodies also produce specialised indices. Some are particularly relevant for the construction industry. For example, the Royal Institution of Chartered Surveyors produces the **Building Cost and Information Service** (BCIS) which is published quarterly. The two main BCIS indices are briefly considered on page 247, but first we need to explain how the most common measures of inflation are calculated.

Cost of Living Indices

In the UK, the most commonly used price indices are the **retail price index** (RPI) and the **consumer price index** (CPI). The two indices are very similar. They both provide an assessment of the prices of goods and services purchased by the typical household. In fact, a good way to think about the RPI and the CPI is to imagine a shopping basket full of goods and services on which people typically spend their money; but in this case the basket includes everything from food and housing to entertainment. Movements in the RPI and the CPI, therefore reflect changes in the cost of living, and they are the accepted measures used to calculate the percentage increase in inflation.

To be precise, the CPI and RPI measures are based on the regular comparison of the prices of about 650 items. Obviously, it is only possible to calculate a meaningful average price for standard products – goods and services that do not vary in quality from month to month or from one retail outlet to another. The prices of these 650 standard items are recorded at various retail outlets in 150 locations throughout the UK. In comprising this monthly index, therefore, nearly 100,000 (150 times 650) prices are recorded; however, only the national averages are published.

An important objective is to make sure that the basket is representative of the consumer goods and services typically purchased by households. Hence, the basket is reviewed each year to keep it as up to date as possible and to reflect changes in buying patterns. Items are dropped from the basket when they become more difficult to find in the shops or no longer typical of what most people spend their money on. For example, purchases of vinyl records were common during the 1960s and 1970s. However, with the advent of compact discs, records now form a niche market and have not been included as a representative item since the early 1990s. Similarly, spending on conventional CRT (cathode ray tube) televisions has decreased and no longer warrants inclusion, and spending on flat-panel televisions has taken their place. It is equally important that new items are added to represent emerging markets, consequently satellite navigation systems, DVD recorders and DAB radios were all added as new items in 2007.

The various items in the RPI and CPI basket are given a statistical weighting to take account of their importance to the typical household. The items that take more of people's incomes are given a higher weighting. For example, the statistical weight for food is higher than that for tobacco and alcohol, as changes in food prices affect everybody, whereas tobacco and alcohol prices only affect smokers and drinkers.

The contents of the baskets used for the RPI and CPI are very similar but not identical. The RPI basket includes several items chosen to represent housing costs, including mortgage interest payments, council tax and rent, all of which are excluded from the CPI. In contrast, the CPI basket includes the fees paid by people living in communal accommodation such as nursing homes, retirement homes and university halls of residence. These so-called institutional households are excluded from the RPI, as its focus is exclusively on private UK households. In addition, the CPI also includes some items to represent costs faced by foreign visitors to the UK. These subtleties form the main technical differences between the two indices.

The more distinct difference between the indices is in their use and origin. The RPI has been established as a means of measuring price inflation in the UK since 1947. The CPI has its origin in a system that is common to the European Union (EU). It was previously called the **harmonised index of consumer prices**, and it allows direct comparisons to be made with the inflation rates in other member states of the EU. It became the official index used by the UK government to target inflation in January 2004. The CPI tends to produce a lower rate of inflation than the RPI, and on average the difference is approximately 0.75 per cent.

In order to distinguish between the two it might be useful to think of a headline rate and an official rate. The **headline inflation rate** is the one that appears in the press and usually refers to the RPI. The **official rate of inflation** forms the target for government purposes and, at present, is the CPI. Finally a passing reference should be made to the retail price index excluding mortgage interest payments (RPIX), as this had been the official rate for government purposes from January 1998 to December 2003. The three measures are compared in Table 14.2.

Table 14.2 Summary of inflation indices

CPI	Inflation measured by the consumer price index, the UK government measure since 2004.
RPI	Inflation measured by the retail price index, in use since 1947.
RPIX	RPI *excluding* mortgage interest payments, was the UK government measure from 1998-2003.

Building Cost Index and Tender Price Index

None of the cost of living indices – CPI, RPI, and RPIX – reflect the prices facing builders and their clients. These indices only indicate the movement in prices of goods and services purchased by the average household, they do not take into account the movement of prices affecting construction. The property and construction consultants Davis Langdon have tracked movements in the input costs of construction work in various sectors since 2000 and these highlight how the rate of increase in construction costs has considerably outstripped the consumer price index. As Figure 14.2 (on page 248) indicates, the construction indices show year-on-year inflation for the period 2000 to 2007 but at a far faster pace than that recorded by the consumer price index. The trend shown in Figure 14.2 has been typical for the past hundred years, as building costs have always increased faster than the general price index – in other words, the real cost of building has risen. This can prove potentially difficult for companies in the construction industry

Figure 14.2 Building cost indices

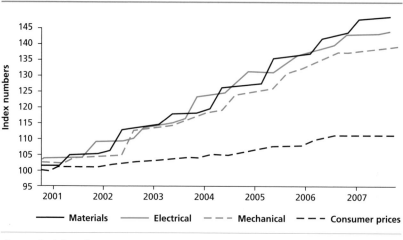

Source: Davis Langdon

because the long duration of large contracts means that work is often let at fixed prices before a project commences.

To assist and support the bidding process, the Building Cost and Information Service publishes a building cost index and tender price index that may be used to monitor changes in construction prices. The building cost index measures changes in costs of labour, materials and plant – that is, it covers the basic costs faced by contractors. The basket is compiled from nationally agreed labour rates and material prices. The index also includes forecasts to help predict any changes in prices that may occur in the period between submitting a tender and project completion.

The tender price index involves an analysis of successful tenders for contracts worth more than £250,000. It includes movement in wage rates, discounts, plant costs, overheads and profit – that is, it indicates the basic cost of construction work to the client. In effect, the tender price index is a measure of the confidence in the industry about its current and future workload. When demand for the industry's services is high, not only do contractors' margins increase but so do the margins of their suppliers and wage rates, and you would expect to see rises in the tender price index. And, conversely, when demand for the industry's services decline, all these factors decrease and thereby exert downward pressure on the price index.

The size of the gap between these two indices suggests something about market conditions. During a recession you would expect to see the indices converge, with tenders certainly at a lower level than during a boom. The explanation is simple: when there is less work available, contractors will be satisfied just to get a contract and cover their costs; when the market is buoyant, tender prices will increase as contractors take advantage of the opportunity to more than cover their costs.

As with retail prices indices, it is important to make sure that in compiling construction indices the prices being recorded are for comparable items. Are apples being compared with apples, not with oranges? In other words are the baskets being used for comparisons consistent? The physical and functional qualities of the structures being priced must be broadly similar to make a valid index.

Key Points 14.1

❑ Inflation is a persistent increase in prices. Controlling inflation has been a main economic priority for UK governments since 1997.

❑ Many modern contracts are index linked to protect the economically active and those on fixed incomes from the effects of inflation.

❑ Price indices compare the current cost (the cost today) with the cost of the same item(s) in a base year.

❑ The RPI and the CPI are both used as measures of the general inflation rate in the UK.

❑ There are several specialised price indices, and examples that apply to the construction sector include the Davis Langdon index, the building cost index and the tender price index.

THE CAUSES OF INFLATION

Although there are many different explanations for inflation, we shall just consider the two main causes. According to these two explanations, inflation either occurs because an increase in demand pulls up prices or because an increase in the cost of production pushes up the price of final products.

Demand-pull Inflation

As we have already explained in Chapter 13 (and reviewed in Key Points 13.3 and 13.4), when aggregate demand in an economy is rising inflation may occur. The severity of the problem depends on how near the economy is to its full capacity level (that is, to full employment). As Figure 13.3 (see page 229) illustrates, there is during any specific time period a fixed output rate in the economy and if demand increases beyond that point the only way businesses can respond is by increasing their prices. Indeed, beyond a point of full capacity no more output is physically possible and the pressures of excess demand can only be countered by price increases. To put it in other words, when total demand in the economy is rising and the capacity available to produce goods is limited, **demand-pull inflation** may occur.

Cost-push Inflation

Prices also rise when the economy is nowhere near full employment. For example, the UK economy experienced inflation during the recession in the mid-1970s. Consequently, other explanations of inflation have been developed. A common feature of these explanations is the focus on changes in business costs caused by factors such as wage rises, widening profit margins or increased import prices for raw materials.

Let's consider oil prices, as they have a history of rising rapidly. For example, oil prices trebled in 1973–1974 following the action by **OPEC** to limit oil production.

They more than doubled in 1978–1979. Some thirty years later a similar saga played out, with crude oil prices rising from around $30 (£17) per barrel at the beginning of 2004 to around $130 (£59) per barrel in May 2008 (Barwell et al. 2007: 522). In each instance, most oil importing countries experienced significant periods of inflation. This is because petrol, propane (LPG), diesel and kerosene (jet fuel) is all derived from crude oil, and oil is also used in the manufacture of many chemicals, plastics, fertilisers and textiles. In short, the lifeblood of any developed economy is dependent on oil. As the price of oil increases, the costs of production are pushed up, which, in turn, are passed on to the consumer in the form of higher prices. This phenomenon is known as **cost-push inflation**.

Two closing reflections on these periods of oil price instability put the problem of cost-push inflation into a sharper focus. First, one response has been to make energy efficiency an important criterion when designing buildings, and experiments based on passive solar design and other renewable technologies have emerged. Second, alongside the search for alternative energy sources, the increasing cost of oil provides the necessary impetus to develop oil fields that previously were not economically viable, such as those below the ocean bed in seas that exceed a depth of 2,000 metres off the shores of Africa and Brazil and those that lie beneath the frozen ice of the Arctic region. Other solutions to inflationary problems are more oriented towards government economic strategy, and some of the policies followed since the 1970s are briefly reviewed next.

CURES FOR INFLATION

Taking a broad chronological overview, three 'cures' for inflation have been deployed in turn since 1970:

- prices and incomes policies
- control of money supply
- interest rate manoeuvres.

Prices and Incomes Policies

During the 1970s, the favoured policy device for stopping, or at least slowing down, cost-push inflation was various versions of wage and price controls. In general terms, these involved employers, unions and governments getting together to agree an annual 'norm' for wage increases. The negotiations, however, proved difficult to manage and, at best, prices and incomes policies only effectively restrained cost-push inflation for a temporary period.

Control of Money Supply

The 1980s witnessed a shift in policy emphasis as some economists argued that inflation, especially demand-pull inflation, seemed to be inextricably related to the size of a nation's money supply. The glib explanation, popularised in the media, was that 'too many pounds were chasing too few goods'. It is difficult to fully appreciate the money supply argument until it is understood that 'money' in a modern society not only comprises notes and coins but also holdings in bank accounts, savings accounts and government bonds. We do not, however, need to be too concerned with the intricacies of this method of control: it has dropped from favour as governments throughout the world experienced too many difficulties in targeting and controlling money supply. In 1977, the OECD published a table showing how the 24 countries that were members of the organisation were using 23 different definitions of money supply. In fact, there were four different definitions of money supply used in the UK during the operation of the policy.

Interest Rate Manoeuvres

Since the 1990s, governments have tried to bring inflation under control using interest rates. This is done on the understanding that the prevailing rate of interest significantly influences spending decisions, and, in particular, affects the decisions of both businesses and households on whether to borrow (that is, to incur debts) to pay for consumption and investment goods. As interest rates become higher, and more volatile, businesses and consumers generally become less confident about making new investments and negotiating future contracts. That is, other things being equal, higher rates of interest should encourage saving and discourage investment and consumer spending. To follow one sequence, higher interest rates will tend to increase the cost of financing house purchases, and so reduce demand and lower (or slow the rate of increase of) house prices. In summary, therefore, changes to interest rates have a strong influence on the level of spending in an economy.

As detailed in Chapter 12, the process of deciding interest rates begins with the monetary policy committee (MPC). The MPC has full responsibility for determining the rate of interest used by the Bank of England when dealing with other financial institutions trying to raise funds in the money market. The official **base rate** is sometimes referred to in monetary circles as the **repo rate** (which represents the rate that a central bank is willing to lend funds to other banks). This phrase relates to repurchase agreements (and sales) of assets such as government bonds between the Bank of England and its counterparties in the money market. The eurozone equivalent set by the European Central Bank is called the **refinancing rate**, and the interest rate used by the Federal Reserve in America is called the **discount rate**. At the introductory level you don't need to worry too much about the difference between the American, British or European terminology: the important point is that in each case the central bank sets a carefully considered unique rate of interest that determines the rate at which it will lend short term to the banking sector. Subsequently this determines the rate of interest that tthe leading banks charge each other for loans. (This rate is sometimes called LIBOR, which is short for the London interbank offered rate.)

In turn dealings in the interbank market affect the interest rates adopted in the wider economy by all other credit agencies, such as the high street banks when trading with their customers. No bank would lend to customers at a lower rate than the one at which it borrows, as it is in business to make a profit not a loss. So, when banks find themselves facing higher funding costs and more difficult economic times, they tend to increase the rates charged on their loans and they might even decide to ration the quantity of credit they are wiling to extend – in effect both these actions were captured in the phrase **credit crunch** that dominated the headlines throughout 2007 and 2008.

CREDIT CRUNCH

The so-called credit crunch emerged from instability in the US financial markets during 2007 and took hold in the UK during 2008. Banks and building societies tightened lending criteria and raised their rates of interest despite the fact that central banks such as the Bank of England had cut their rates several times. This action was unprecedented as the usual practice (outlined above) is for bank interest rates to mirror adjustments and certainly to move in exactly the same direction as changes made to the official (repo) rate – and usually on the same day. During 2008 however, institutions in the financial market were less quick or symmetric in their rate changes as they began to revise their margins and increase the costs of borrowing to move to a new market equilibrium. In short, the crisis captured by the phrase credit crunch represented a period of time when global financial markets sought correction.

The important point is that the official rate (known as the base rate or repo rate) is the reference point that ultimately affects the whole spectrum of interest rates used across the economy. The sequence of events is neatly captured in Figure 14.3. In general, it can be assumed that the rate of interest always changes in marginal increments as we progress from the Bank of England to the ordinary customers of commercial banks. The credit crunch dimension simply means that the gap between the official repo rate and all other rates significantly widened; highlighting that, in relative terms, borrowing costs had increased and savings rates had decreased.

Figure 14.3 The money market

In economic terms the credit crunch has serious implications, as businesses and individuals have to adapt to the new market rates of interest on their investments, savings and debts. Large firms feeling the pinch will tighten up the contractual terms on their supply chain and smaller firms often do not survive. In January 2008, for example, it was widely reported that Taylor Wimpey, the UK's largest housebuilder, sent a letter to firms in its supply chain stating that due to economic circumstances it needed to take urgent action and impose a 5 per cent cut on all existing and future contracts (Richardson 2008). As a result of the credit crunch, many small and medium-sized firms became insolvent and the impacts were strongly felt in the construction sector: 76 construction companies went bankrupt in the month of March 2008 and a further 2,000 were forecast to follow before the year was out (Bill 2008).

It takes time for this type of economic instability to play out. There is a level of uncertainty about the impact of interest rate changes. According to the Bank of England macroeconomic model, interest rate changes can take up to one year to affect demand and output, and nearly two years for all the effects to be reflected in the inflation rate (MPC 1999: 3). It is quite possible that several months may pass before a higher rate of interest impacts on the payments made by mortgage holders, especially as those on fixed rate mortgages will not face higher charges until their fixed term expires, and it will be even longer before this feeds through to changes in retail spending. Subsequently, there will be cut backs in the quantity of orders made to distributors, manufacturers and other producers. As a consequence, employment and earning levels will reduce and, through a multiplier effect, this will then feed into further reductions in domestic demand.

The interest rate approach to controlling inflation explicitly rejects the old ideas that price stability can only be achieved by imposing legislative constraints on wages and prices or by managing the long-run level of aggregate demand through control of the money supply. Instead, the current framework is based on the conviction that the best contribution that monetary policy can make to long-term growth and employment is to deliver price stability by managing the spectrum of interest rates. In the next two sections, we broaden out the theoretical economic arguments that explain the importance of a low and stable inflation rate to business.

Key Points 14.2

- ❑ Demand-pull inflation occurs when the total demand for goods and services rises faster than the rate of growth of supply.

- ❑ Cost-push inflation is due to one or more of the following: (a) wage rises, (b) widening profit margins, and/or (c) raw materials price increases.

- ❑ Prices and incomes policies, control of money supply and managing interest rates have all been used as attempts to reduce inflationary pressures.

- ❑ The credit crunch represents a period of tightening lending conditions, both in terms of cost and availability. A central bank cannot simply lower interest rates to solve this problem as inflationary pressures must be kept in check.

EXPECTATIONS

Today, economists and politicians believe that expectations influence all types of economic variables including inflation. In the past, these expectations had been considered as external influences that were beyond control. In formal terms, they represented an example of an exogenous variable. Modern economic theory, however, seeks to understand behaviour that is based on expectations. Initially, during the 1970s, the **adaptive-expectations hypothesis** was accepted.

Adaptive Expectations

The basic idea behind this theory is that people take what happened in the past as the best indicator of what will happen in the future. Look at Figure 14.4. Here we represent the actual rate of inflation by a black dashed line. Inflation is rising: in year one it is 6 per cent, during the second year it is 8 per cent, and in the third year it jumps to 12 per cent; thereafter it stays at 12 per cent.

According to the theory of adaptive expectations, the expected rate of inflation in any given year is whatever the rate of inflation was in the previous year. The orange line in Figure 14.4 represents this expected inflation rate. During the second year, people believed inflation would be 6 per cent because that is what it was the year before. During the third year, they believed it would be 8 per cent (because that is what it was the year before). In this simplified model, people's expectations are always behind reality. When the rate of inflation is rising, they will always believe

Figure 14.4 Adaptive expectations theory

According to this hypothesis, individuals formulate their expectation of inflation according to the rate in the previous year. In the figure, the actual rate of inflation is shown in black and the expected rate of inflation is shown in orange – and it clearly lags one year behind in pattern. For example, in year 2 people expect inflation to be 6% because that is what it actually was in year one.

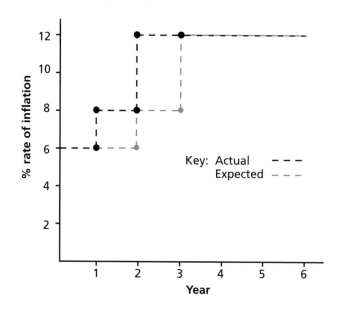

that inflation is going to be less than it actually turns out to be; when the rate of inflation is falling, they will always believe it is going to be more than it actually turns out to be. Only when the rate of inflation remains constant for a sustained period of time do expectations come into line with reality.

A more sophisticated model of expectations would not allow people to be consistently proved wrong. It would recognise that opinions are based on a range data, which includes historical trends but not to the exclusion of everything else.

Rational Expectations

The **rational-expectations hypothesis** is the logical development of the adaptive-expectations model. It is more concerned with human behaviour and it allows for the sifting and weighing of all available information. In this theoretical framework, people do not simply look at what has happened in previous years. They consider what has happened in the past and combine it with their knowledge of current government policies and any anticipated policy changes. Rational expectations, therefore, require judgement about current and future policies combined with lessons from the past. For example, the shrewd forecaster considering house prices in 2006 would recognise that prices do not continue to rise forever – and that the rises in previous years could not necessarily be sustained.

Rational expectations take on more importance in some sectors of the economy than others. For example, city brokers, investment managers, property developers, fund managers, estate agents and those involved in other valuations depend upon their specialist skilled knowledge to make shrewd judgements. It is debatable, however, whether managers, workers and consumers in the general economy can be expected to have developed the same level of expertise. Consequently, the general public often adopts the expectations of the professionals. This may be better than the 'blind leading the blind', but it can lead to scenarios where power psychology seems to generate interesting outcomes. Two detailed examples are considered next.

THE WAGE-PRICE SPIRAL

Expectations can be regarded as the driving force that underlies the beginning and end of a wage-price spiral. If we consider Figure 14.5 (on page 256), it is evident that, regardless of the initial cause of inflation, workers eventually obtain a wage increase which subsequently leads to prices being marked up. The interesting point, however, is that what perpetuates and accelerates this spiral-type behaviour is expectations; most notably the expectations of those in charge of governments, trade unions and employer confederations. In turn these expectations are exacerbated further by the media coverage of inflation (Blanchflower 2008: 11).

Consider the position at box 5 in Figure 14.5. At this stage in the proceedings, it is most likely that labour correctly anticipates further inflation. Workers not only negotiate a wage increase to restore their purchasing power, but they also seek to incorporate a 'hedge' against future inflation. For example, if there has been a period of inflation at 15 per cent, workers will demand a greater than inflation pay increase of, say, 30 per cent. This would allow 15 per cent for the previous inflation, 10 per cent for anticipated inflation, and 5 per cent which can be given up during

Figure 14.5 The wage-price spiral

negotiations. Next, we can see – in box 6 – that entrepreneurs employ a similar rationale and increase prices sufficiently to cover the total increase in their wage bill plus a precautionary amount to avoid having to make further price increases too soon. In both cases, the wage claims and price increases are being made on the basis of judgements about the future. By box 7, therefore, it should be evident that judgements anticipating inflation become firmly entrenched and create a self-fulfilling prophecy; with workers seeking compensation for rising prices through higher wages and companies attempting to pass on cost increases to their customers. In short, it is rational expectations that drive the wage-price spiral.

Since the late 1980s, governments have tried to break the mentality associated with the wage-price spiral by convincing businesses, workers and the general public that their policies should slowly eradicate inflation, thereby encouraging people to lower their judgements about the future inflation rates. Indeed, this is why governments in countries such as New Zealand, Canada, Sweden and the UK have moved towards the idea of explicitly targeting inflation. When Balls and O'Donnell (2002: 72) were economic advisers to the government, they noted that an explicit target for inflation is desirable as 'it provides a clear anchor for inflation expectations, making it easier to attain and maintain price stability'. Consequently, whenever the trend of low inflation shows signs of changing course, government ministers carefully account for these 'blips' in the monthly CPI figures and address people's worries as expectations must not be allowed to change. The problem with economic forecasting, however, is that unforeseen events occur, 'blips' can turn into 'blots', and prices can start to rise again. Once this occurs, rational expectations will see the wage-price spiral accelerate upwards with a vengeance. This underlines the fact that the psychology of expectations plays a crucial part in determining all economic destiny, especially that of inflation.

TIME-LAG

A similar type of analysis can be applied to those market sectors in which the nature of the product means that a long time elapses before changes in output can be realised. The markets normally considered in this context are agriculture and construction. The problem is that the inflexible nature of these markets causes them to respond slowly to price changes. Furthermore, these markets are often structured in such a way that factors such as trade union practices, training requirements, information flows and legislation also restrict the dynamics of the market. Economists refer to these problems as structural rigidities.

Construction markets – as the cyclical booms and slumps demonstrate – struggle to arrive at stable market prices. The cobweb theorem can be used to explain the circumstances. As we described in Chapter 6, construction can be a lengthy and fragmented process. For example, it takes approximately 18 months to two years to complete a house from the initial planning stages and up to five years to deliver a modern office block. As a result, price instability within the construction sector and its associated markets is quite common. As an example, we show the percentage change in house prices according to three sets of indices in Figure 14.6. Though there is some divergence between them, the three indices each illustrate the key point that house prices tend to be unstable.

To account for this instability, we need to accept two distinguishing features of the construction industry.

- The decision of builders to change their output is heavily influenced by current market prices. So, for example, house supplies in two years time are dependent on today's prices.
- There are many small building firms and they each take decisions to adjust their scale of production in isolation from each other.

Figure 14.6 UK house price inflation

The graph shows the unstable nature of prices between 1995 and 2006. The cobweb theorem, illustrated in Figure 14.7, helps to explain these price fluctuations.

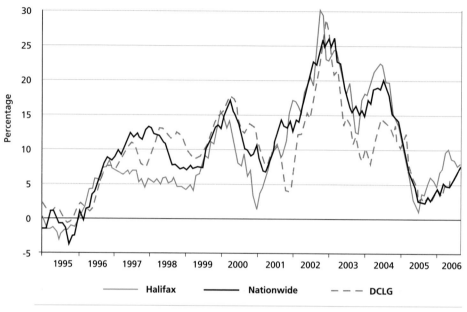

Source: ELMR Jan 2007

If buildings happen to be scarce in one time period, then high market-clearing prices will prompt construction firms to begin new projects. This rise in production will in due course depress market prices, which sets off a major contraction in the scale of production. We show this process in Figure 14.7. The fixed amount of property available in the short run means that the market-clearing price may rise to P_1. The supply curve indicates that, at price P_1, construction firms would like to produce Q_2, although they cannot do so immediately. Building contractors begin to make plans to adjust output based on the current market price and try to produce Q_2. After a time-lag, a potentially unstable situation is likely. Indeed, assuming the pattern of demand remains unchanged, and output Q_2 is actually built, prices will have to drop to absorb the excess supply. This will ultimately result in a contraction of production, leading once more to higher prices. Note that in this example, market prices move above and below the equilibrium level and are not stable. This price instability could be greater or less depending on the elasticity of the supply and demand curves.

The pattern of oscillation around the equilibrium point is accounted for by the inherent time-lag in producing construction output, by imperfect information and, to a lesser extent, by the inflexible nature of the construction market. The oscillating pattern also provides the explanation for the name of this model – the cobweb theorem.

Figure 14.7 A cobweb diagram showing how property prices can fluctuate

Commencing from the equilibrium price, a change in the conditions will move the market from its equilibrium position. In the short term, the market is faced with a perfectly inelastic supply curve, as shown by the dashed line (P_{ES}). The longer-run supply curve S indicates that at price P_1, construction firms would like to build Q_2. These structures will take some time to construct. However, when Q_2 is eventually completed, the price will need to fall to P_2 to absorb the excess supply. This will result in a subsequent contraction of supply as shown by position 3. The shortage of supply will cause prices to be bid up and the process continues.

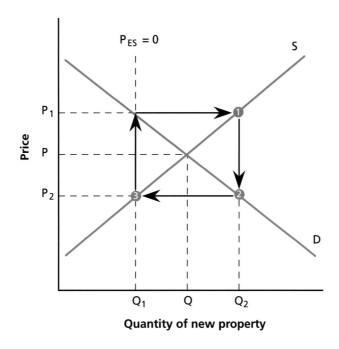

Quantity of new property

Obviously, there are limitations to this generalised theoretical model. First, builders are aware of the cyclical nature of their industry and attempt to make adjustments to relate starts to expected prices rather than current prices. Second, they can choose to manipulate their stocks by holding property off the market during the low price part of the cycle. Nonetheless, the economic analysis is still useful since it highlights the root of a problem, and amplifies the need for a stable economy to secure the necessary business confidence for construction firms to maintain a steady rate of production.

Key Points 14.3

❏ The adaptive-expectations hypothesis argues that people make predictions about the current year's rate of inflation based on the previous year's rate of inflation.

❏ The rational-expectations hypothesis assumes that individuals form judgements by examining all available information. That means that economic forecasters not only look at past data and trends but also at the impact of current and/or anticipated future government policy.

❏ The wage-price spiral represents the inflationary process as one in which incomes and prices continually chase each other in an upward direction (see Figure 14.5).

❏ Construction firms operate in a different business environment as significant time-lags exist before any planned changes in output can be realised. These production time-lags together with several structural rigidities cause construction markets to be unstable.

❏ It is rational expectations that determine the speed and direction of price changes.

An important theme in Part C has been the idea that construction plays a leading role in the process of national development and macroeconomic management. The following two extracts analyse the shape of the Bon curve, which is derived by plotting GDP per capita against expenditure on construction to show the importance of this sector to economic growth. The first edited extract is based on the empirical evidence from 39 countries, at different stages of development, during 1994–2000. The authors adopt a clear and general approach to introduce and interrogate Bon's theory. The second extract is slightly more advanced, in the sense that it seeks to question the exact shape of the curve for smaller nation states, such as Hong Kong, Singapore, and Trinidad and Tobago, and provides a case study approach that focuses on data relating to Cyprus during 1998–2005. Some obvious questions are suggested by studying these extracts. For example, does the data reflect the current relationship between construction output and GDP in your country? In other words, use recent construction data to clarify your understanding of the sector's broader impacts on the current pace and direction of economic growth. What do you think of the closing idea posed in the second extract that micro-states with a fragile ecosystems will inevitably experience sharper limits on construction growth than larger scale economies? Finally, what are the implications of the dynamic role of construction for macroeconomic management? In particular, consider the importance of government policies designed to secure growth, maintain stable prices and protect the environment.

Les Ruddock and Jorges Lopes (2006) 'The construction sector and economic development: the Bon curve' *Construction Management and Economics* 24: 717–23

Introduction

The relationship between a country's state of development and the level of activity in the construction sector is one, which has been the subject of study at the macroeconomic level for a number of years (Turin 1973; World Bank 1984; Wells 1987; Bon 1990). A major obstacle to such studies has been the lack of appropriate information on the sector, particularly in developing countries....

Bon (1992) analysed the changing role of the construction sector at various stages of economic development and presented a development pattern for the industry based on the stage of development of a country's economy. This notion is well-explained in basic terms by Tan (2002: 593): "In low income countries, construction output is low. As industrialization proceeds, factories, offices, infrastructure and houses are required, and construction as a percentage of gross domestic product reaches a peak in middle income countries. It then tapers off as the infrastructure becomes more developed and housing shortages are less severe or are eliminated." An important aspect of the proposition was that, in the early stages of development, the share of construction increases but ultimately declines, in relative terms, in industrially advanced countries – and even at some stage, the decline is not only relative but also in absolute terms, that is 'volume follows share'......

In Bon's 1992 paper, the link between economic development and construction is discussed; his basic premise is that the share of construction in total output first goes up and then comes down with economic

Figure 1 Bon curve

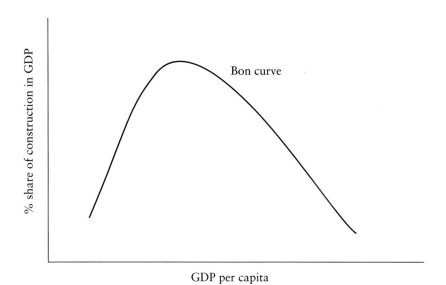

development, and this is called the inverted U-shaped relationship plotted in Figure 1.... His paper concerned the entire path, from the least developed countries (LDC) through to those that are classified as advanced industrialized countries (AIC).....

Data

The indicator of construction industry activity used for this analysis is gross value added (GVA) in construction.

GVA in construction is calculated the same way as in any other sector, but includes only the activities of the construction activity proper. For example, it excludes the building materials industry which is accounted in the manufacturing sector. The main indicator of general economic activity used in this study is GDP per capita. It adjusts the growth in the economy with the growth in population. It is a better indicator of a country's welfare particularly in developing nations, where the growth rate of population has been since the Second World War roughly twice as high as in developed economies.....

Does the volume follow the share?

The analysis depicted clearly demonstrates that the inverse 'U'-shaped pattern holds for the share of construction in the national economy. That is the share of construction in

total output first goes up and then comes down with economic development. Table 1 shows Bon's grouping of countries, according to their status of development, categories 1 and 2 taken together represent the LDCs (and those that are emerging such as India and China), category 3 the newly industrialised countries (NICs) and category 4 represents the AICs.

What about Bon's proposition that at some stage the decline of the construction industry activity switches from relative to absolute decline, that 'volume follows share'? Looking at Table 1, it can be seen that the share of construction in the advanced industrial countries increased by 6.5% in the period 1994–2000, which suggests that construction volume did not decrease in these countries over the period. This is consistent with Carassus' (2004) findings that the construction volume, measured in constant prices, in seven out of eight advanced industrial countries did not decline in the period 1990–2001. OECD (1998) also shed new light on the argument, as data for the European Union 15 shows that both GDP and gross fixed capital formation (GFCF) increased in the period 1960–1996, though remained generally stagnant in the period 1974–1982. The data also shows that the growth in GDP is higher than that of GFCF, thus the relative decline in capital formation is also evident for the entire period.....

We do believe that Bon's (1990) analysis which concerned the period 1970–1985 was certainly influenced by the recessive period 1973–1982, characterized by two oil shocks, that followed the period 1960–1973 – the 'golden age' of the world economy...

Table 1 Relative growth rate in construction (1994-2000)

GDP per capita in 1994 ($US)	LDC		NIC	AIC
	<1000	1000 to 2499	2500 to 9999	10000 or over
Relative change in construction (%)	-5.0	11.1	10.7	6.5

Note: The figure shows the average for the countries in each category

References

Bon, R. (1990) The World Construction Market 1970–85, in *Building Economics & Construction Management: Proceedings of the CIB W65 Symposium, Sydney.*

Bon, R. (1992) The future of international construction: secular patterns of growth and decline. *Habitat International,* 16(3), 119–28.

Carassus, J. (2004) *The Construction Sector System Approach: An International Framework,* CIB: Publication 293, CIB, Rotterdam.

OECD (1998) *National Accounts: Main Aggregates 1960–1996, Vol.1,* Organisation for Economic Co-operation and Development, Paris.

Turin, D.A. (1973) *The Construction Industry: Its Economic Significance and its Role in Development,* UCERG, London.

Wells, J. (1987) *The Construction Industry in Developing Countries: Alternative Strategies for Development,* Croom Helm Ltd, London.

World Bank (1984) *The Construction Industry: Issues and Strategies in Developing Countries,* The World Bank, Washington, DC.

Extract information: Edited and adapted from pages 717–18 and 721–2, plus relevant references from page 723.

Second part of Reading 5 continues overleaf

Ozay Mehmet and Vedat Yorucu (2008) 'Explosive construction in a micro-state: environmental limit and the Bon curve: evidence from North Cyprus', *Construction Management and Economics* 26: 79–88

The shape of the Bon curve

Although the existence of the Bon curve [see Figure 1, page 262] enjoys general acceptance, its shape is not yet settled. The exact nature of the construction–GDP relationship is a matter of empirical question and several questions of theory emerge over stages and transformations, over and above problems of statistics, sources and methodology. Thus, some question the slope of the Bon curve during the drive to, and beyond the maturity stage. Accordingly, as economies 'mature', construction activity slows down and the declining portion of the Bon curve becomes flatter, indicating a declining importance of the construction sector in the economy....

Structural transformation within the construction sector

Within the aggregate relationship revealed by the Bon curve, there are important transformations because the construction sector is a diverse sector, consisting of residential, institutional, commercial, industrial, infrastructure, maintenance and other subdivisions. During the course of economic development, the relative significance of these construction sub-sectors are altered in tandem with the Colin–Clark hypothesis (Mehmet 1999: 96) that with a rising GDP, the structure of the economy changes from being agrarian to industrial and then to service-based economic activity. Pietroforte and Tangerined (1999), using disaggregated Italian data for the 50 largest construction firms, have demonstrated that during the 1980s and 1990s these firms have responded to changing industrial policy to adapt a service orientation in construction activity in order to remain competitive. One finds Clark-type evidence in earlier construction activity research as well. Thus, Strassman (1970) argues that the construction sector, like agriculture or manufacturing, follows a pattern of change that reflects a country's level of development. After lagging in early development, construction accelerates in middle income countries and then falls off. The extent of construction activity in an economy is closely linked in particular to the extent of urbanization in the economy. The share of urban population increases initially and then decreases as GDP per capita rises.

Does size affect the Bon curve?

The importance of the size of an economy has traditionally been ignored. Generally smallness has been considered to be a negative, limiting factor on economic development....But this is misleading as some of the most prosperous countries in the world are micro-states, with less than 10,000 km sq, such as Singapore, Cyprus, Trinindad and Tobago, Hong Kong, etc. (Mehmet and Tahiroglu, 2002). At the same time, however, construction activity may be more volatile in micro-states. Thus one major project coming to an end can have a major impact on the aggregate data, whereas in larger economies any single project will only form a relatively small proportion of total construction output. This may exhibit greater short-term volatility than the large countries. In any event, construction activity in micro-states is an important topic in its own right given the fact that some of the wealthiest countries in the world are micro-states....Accordingly, it is not an idle question to inquire whether or not construction in small states contradicts or follows the shape of the Bon curve....

In Trinidad and Tobago Ramsaran and Hosein (2006) found that construction has become very much a leading sector of the economy, manifesting a powerful multiplier impact in both boom and bust periods, growing at twice the GDP rate and that the contribution of construction to economic development was positively correlated with the domestic resource content of construction output. This is the pattern observed in North Cyprus as well (Yorucu and Keles, 2007) where small size is encountering a critical ecological limit which larger states may be better able to postpone.

The Bon curve and the environmental limit — a hypothesis from North Cyprus

The other major consequence of smallness is the prospect of an earlier environmental limit than in larger states. Concern for the protection of a fragile ecosystem, watershed or natural heritage becomes evident as income per capita rises, and when explosive construction-led growth occurs, as in the case of North Cyprus after 2003, public demand for tougher environmental standards may lead to a sharper downturn in construction than in larger states. Theoretically, this implies that in micro-states the declining portion of the Bon curve may be steeper than in larger economies. In other words for micro-states like North Cyprus, an inverted V–shape may be more appropriate than an inverted U–shape and this is portrayed in Figure 2. In micro-states with land constraint and access to cheap labour supply, the expansion portion of the Bon curve may be steeper, the peak much more pronounced and the declining portion relatively flatter compared to the traditional Bon curve....

Figure 2 Bon curve

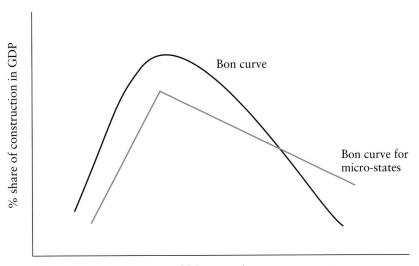

References

Mehmet, O. (1999) *Westernizing the Third World, Eurocentricity of Economic Development Theories*, Routledge, London and New York.

Mehmet, O. and Tahiroglu, M. (2002) Growth and equity in microstates: does size matter in development? *International Journal of Social Economics*, 29(1–2), 152–62.

Pietroforte, R. and Tangerined, P. (1999) From boom to bust: the case of Italian construction firms. *Construction Management and Economics*, 17, 419–25.

Ramsaran, R. and Hosein, R. (2006) Growth, employment and the construction industry in Trinidad and Tobago. *Construction Management and Economics*, 24, 465–74.

Strassmann, W.P. (1970) The construction sector in economic development. Scottish *Journal of Political Economy*, 17(3), 390–410.

Yorucu, V. and Keles, R. (2007) The construction boom and environmental protection in Northern Cyprus as a consequence of the Annan Plan. *Construction Management and Economics*, 25(1), 77–86.

Extract information: Adapted from pages 80–2 plus relevant references from page 87.

15 Sustainable Construction

Throughout this book we have tried to capture the main ideas that underpin the concept of sustainable construction. In this final chapter, we revisit some of the central issues that shape the sustainable construction agenda. At the outset, it is important to remove any lingering confusions caused by the term 'sustainability'. Sometimes, sustainability is used solely to refer to concerns surrounding the natural environment; at other times, it seems to have a broader connotation, including two other integral strands – sustainable communities and sustainable businesses. The narrow environmental focus is perhaps understandable – both for historical reasons and because, in discussing the construction industry, environmental impacts are significant. However, in this book, we have tried to stress that sustainability is formed of three constituent parts – the community, business and the environment – and this is the focus of the chapter.

SUSTAINABLE DEVELOPMENT

Sustainability is often misinterpreted as being synonymous with the terms 'green' or 'environmental'. However, as we explained in Chapter 2 (see, for example, Key Points 2.4) and reiterated above, this only represents a third of the agenda. It is impossible to talk of one strand of the sustainable development agenda without considering the others – especially if the agenda is to become more than a theoretical construct. Indeed, it is impossible to make an environmental decision without there being economic and social implications.

A useful starting point, therefore, comes from confirmation that sustainability embraces the three themes of environmental, social and economic accountability – often known as the triple bottom line. Sustainability can be achieved by minimising negative impacts and maximising benefits. The best way of doing this is to look for solutions that solve more than one problem at a time. These solutions are known as 'win-win-win', as they secure economic, social and environmental benefits simultaneously. Win-win solutions and triple bottom lines are an ideal way of thinking about sustainability as they emphasise the need to integrate social, environmental and economic issues.

A brief chronology of the development movement shows how the broader concept of sustainability has emerged. During the post-war period in the 1950s and 1960s, development was mainly discussed in relation to the less developed countries. Up until the 1970s, therefore, development thinking was concerned primarily with economic growth and its distribution. A common reference point from this period was Rostow's model of economic growth (see Figure 15.1 on page 268) which portrays development as a set of five linear stages. From this perspective, a key policy aim was to assist the less developed countries through aid and technology transfer to help them reach the critical 'take off' stage. In Chapter 13, we discussed some of the ideas which underpin this model of economic development. For example, we noted that investment in construction is especially important during the development phases of an economy (see, for example, Key Points 13.1).

Figure 15.1 Rostow's stages of economic growth

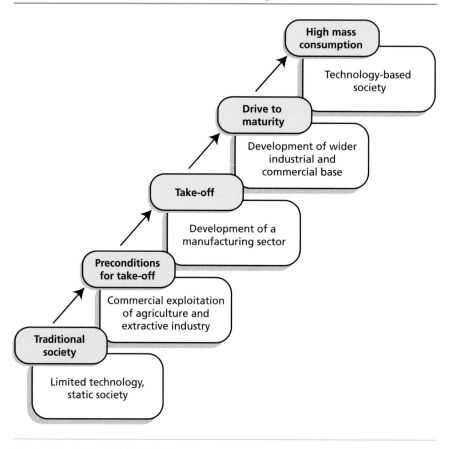

Source: Adapted from Elliot (1999: 11)

The origins of the sustainable development movement dates from the 1970s. In this period, the environmental concerns of the world's poor – such as a lack of clean water or sanitation – began to take centre stage. In 1973, for example, the United Nation's Environment Programme (UNEP) was established to meet demands for an international environmental watchdog. In short, issues relating to the environment and development began to be viewed as interdependent.

The next milestone was the 1987 report of the World Commission on Environment and Development. Entitled *Our Common Future* (WCED 1987) – though more commonly referred to as the Bruntland report after the commission's chair, the former Prime Minister of Norway, Gro Harlem Bruntland – this report was translated into 24 languages and popularised the idea of sustainable development as being concerned with intergenerational equity and justice.

Perhaps the most well-known definition of **sustainable development** comes from the Bruntland report (see the second definition in Table 15.1). However, there are more than 70 definitions – Pezzey (1989) lists 60 alone – and Table 15.1, which lists just three interpretations, gives some idea of the breadth of the concept. When first drafted these definitions were sufficiently vague to allow for a broad consensus, but over time they have become a breeding ground for disagreement.

Table 15.1 Three interpretations of sustainable development

1	It is indistinguishable from the total development of society (Barbier 1987: 103)
2	Development that meets the needs of the present without compromising the ability of future generations to meet their own needs (WCED 1987: 43)
3	Development that delivers basic environmental, social and economic services to all residences of a community without threatening the viability of natural, built and social systems (ICLEI 1996)

It is perhaps easier to clarify the definition by starting at the other extreme and setting out features of 'unsustainability'. Unsustainable development is associated with ozone depletion, poor sanitation, extinction of species and habitat, social conflict, toxic pollution and resource depletion. The origins of the sustainability movement is a reaction to these problems and stems from concerns about the future capacity of the planet's life support systems. Indeed, since the 1970s academics have discussed the problems of prolonged economic growth pushing society beyond global limits – see, for example, any of the titles by Meadows *et al*. The movement gained some impetus from the ideas put forward by environmental economists, including work on the monetary value of externalities and the integration of economic and environmental systems. These ideas were outlined in Chapter 11, and it may be useful here to review Key Points 11.1 and 11.3. In Table 15.2, we contrast unsustainable development with modern ideas of sustainable development.

Table 15.2 What makes sustainable development different?

Unsustainable Development	Sustainable Development
Aims to raise the standard of living, based solely on monetary measurements of gross domestic product.	Aims to improve the general quality of life, including non-monetary factors to do with the environment and community.
Treats the economy, society and the environment as three separate issues.	Sees economic, social and environmental issues as interlinked.
Focuses on improving things in the short term. Mainly leaves issues to do with the future up to those who will live in it.	Looks at the needs of future generations as well as people today, and seeks to avoid problems in the future by taking precautions today.
Treats the environment as a luxury to be protected if we can afford it.	Takes account of the environment and its capacity to support human activity.

Most people fully support the concept of sustainable development, as the associated ideas are not contentious. The challenge lies in finding a way that society can develop sustainably. Throughout this book we have alluded to some signs of progress towards this goal – but we have also made it clear that there is much uncertainty about the most appropriate strategies. For example, in Chapter 11 we distinguished between the approach of neoclassical and environmental economists. The neoclassical school of thought promotes a free market approach, arguing that this should ensure that the depletion of natural resources will be countered by increases in man-made equivalents. According to this technocratic perspective, aggregate capital is kept intact over time. In contrast, environmental economists believe it is necessary to maintain a critical minimum level of natural capital – for more on this, review Key Points 11.1.

In 1992, the United Nations Conference on Environment and Development – the so-called Earth Summit – took place in Rio de Janeiro, Brazil. It was the largest international conference of any kind, with 118 heads of state in attendance. This was clearly a sign of the times, as societies felt threatened by the looming problems of resource depletion and environmental degradation. The main document signed at Rio contained 27 principles for sustainable development and an action plan known as Agenda 21. It was here that the three strands – environmental, economic and social – explicitly emerged.

Rio helped to dispel any idea of achieving environmental objectives without securing the interest and motivation of business and communities. Instead, modern policies of sustainable development highlight this interrelationship, For example, during his time as principal economist at the World Bank, Ken Gwilliam (2000: 368) was keen to emphasise that environmental objectives could not be sensibly pursued in isolation from economic and social criteria. This synergy between environmental, economic and social factors is summarised in Figure 15.2. However, it can also be argued that there are a number of conflicting tensions that limit the synergy between the three strands. Is it really possible to achieve continuous economic growth and manage emissions and pollution? Are opportunities to enjoy the countryside and its associated biodiversity equally available to all sectors of society? Do the fiscal burdens imposed by market-based instruments cause greater cost to the poor or do they apply fairly to everyone? The issues raised by these type of questions are symbolised by the trade-off zones in Figure 15.2.

In the years following the Rio Earth Summit, individual countries adopted different approaches reflecting their own specific priorities. In the UK the principles of sustainable development have evolved over several years. The current policy, as set out in the *UK Strategy for Sustainable Development* (Defra 2005), seeks to deliver:

- prudent use of natural resources to secure sustainable consumption and production
- protection and enhancement of the natural environment
- strong and stable economy to provide prosperity and opportunities for all
- good governance based on sound scientific evidence.

These basic principles have a strong resonance with the three-part framework adopted for this book.

Figure 15.2 The three strands of sustainability

SUSTAINABLE CONSTRUCTION

In several countries, construction has been identified as the first sector to require specific attention in meeting the sustainable agenda. There are several reasons for this 'accolade'. First, in sustainable development terms, construction is consistently responsible for some of the most profound negative impacts. Examples of the industry's use of large amounts of materials and resources, and its reputation as a huge generator of waste were detailed in Chapter 9, in which we reviewed each stage of the life cycle pertaining to the construction process. We noted that the construction industry consumes more raw materials than any other industrial sector and is responsible for a significant proportion of Europe's waste stream (see, for example, Key Points 9.3). In Great Britain alone, it is estimated that 100 million tonnes of construction and demolition materials end up as waste every year, of which 15–20 per cent comprises material delivered to sites and then thrown away unused. In 2006, regulators at the Environment Agency stated that the construction industry was responsible for more than 150 water pollution incidents. Statistics estimating the negative impacts of construction are given in many publications – for example, see Environment Agency (2007), EU (2007) and HM Government (2008).

Second, the construction industry is a vitally important industry. This has already been reviewed at some length. In Chapter 13, we discussed how construction is commonly regarded as the engine of economic growth and illustrated the point with case studies from China, Singapore and Japan, and in Reading 5 we examined the idea further by exploring Bon's (1992) hypothesis that a country's level of construction activity is closely related to its stage of economic development. In purely quantitative terms, the contributions made by construction to employment and GDP are quite significant. In Great Britain, the narrowly defined construction industry employs around 1.5 million people across more than 186,000 firms and the turnover accounts for about 7 per cent of GDP. In Europe as a whole, construction employs more than any other industrial sector – accounting for around 15 million jobs and around 10 per cent of GDP. The industry is not just important economically, it is the key to the quality of life as it produces the built environment and puts in place the physical facilities and infrastructure that determine the degree of freedom and flexibility that society may enjoy. Its products also have a long lifetime, typically for anything up to one hundred years after construction. This puts it in a very different league to, say, producers of photocopier paper, washing machines and cars.

Finally, the most worrying reason that construction has been selected as warranting a special case in the sustainability agenda is because of its perceived lack of change. In nearly every other sector of the economy, technological developments have fuelled changes in business attitudes. For example, the manufacturing industry has become leaner, cleaner and quicker at all tasks. However, most of the construction process continues to be unsustainable and the industry does not seem likely to make any changes on its own accord. As Sir John Fairclough (2002: 30) confirmed in the recommendations to his report, construction is presently 'perceived as dirty, dangerous and old fashioned', and to address this problem it needs to be seen as central to achieving a sustainable future. As pointed out in Chapter 6 similar sentiments are echoed throughout the world. To take a European example, a review of 30 market areas with high economic and societal value identified construction as requiring urgent and coordinated action. To paraphrase the report, the construction industry at national levels across the EU can be singled out as lacking innovation. It is a market of few large players and many small businesses. The owners of these businesses look for job opportunities in their local areas and are not inclined to invest time or money into research and development. Their main concern is to ensure order books for the next 6 to 12 months. This economic reality has an impact on the effectiveness of innovation, business strategy, design activities and training requirements. In other words, the industry of the twenty-first century is still made up of a high proportion of small contractors, who execute business plans across short time horizons, and persist with traditional methods of work (EU 2007: 10).

The notion of sustainability therefore has a special relevance to construction and a specific agenda is evolving. The first international conference on sustainable construction was held in 1994. The conference defined sustainable construction as:

the creation and operation of a healthy built environment based on
ecological principles and resource efficiency.

These principles of sustainable construction have been taken to mean different
things in different countries. Table 15.3 lists the countries that have developed a
specific sustainable construction policy, and in most instances these policies refer to
the narrowly defined construction industry. In some cases, the policies include
economic, environmental and social considerations within their framework and
these are referred to as a 'three strand policy'; others only identify an environmental
strand. In practice, the main impacts of these policies are limited to the narrowly
defined construction industry and generally tend to favour environmental aspects.
All the countries listed in Table 15.3 are in the developed world, but discussions are
also beginning on a separate agenda for sustainable construction in developing
countries where there are significant differences in priorities, skill levels and capacity
(see Du Plessis 2002 and 2005).

Table 15.3 Countries following a sustainable construction agenda

Three strand policy	Environmental strand
Denmark	Australia
Finland	Canada
France	Germany
Ireland	USA
Japan	
Netherlands	
Norway	
Portugal	
Sweden	
United Kingdom	

Sources: Adapted from Manseau and Seaden (2001), EU (2007) and Kibert (2007)

Three interpretations of what sustainable construction involves in practice are
shown in Table 15.4 (0n page 274). The key ideas that recur in these definitions are
to minimise the amount of energy and resources used in the construction process,
reduce the amount of waste and pollution, and to respect the various stakeholders –
particularly the users – both now and in the future. This last aspect seems to be a
particular problem in construction where the client and contractor are so often
separate and indifferent to one another's needs. It seems obvious that an efficient
building is one that supports the client's needs, yet in Chapter 6 we noted the
problematic nature of the relationship between clients and contractors. It is not
surprising that research has shown that well-designed and well-constructed hospitals
aid the recovery of patients, that good schools are those that are designed with
learning in mind. As an architect would describe it, in any building form and
function should be complementary.

Table 15.4 Three interpretations of sustainable construction

In Finland, since 1998, it means

✓ Intensified energy efficiency & extensive utilisation of renewable energy resources
✓ Increasing the sense of wellbeing over a prolonged service life
✓ Saving of natural resources and promotion of the use of by–products
✓ Reducing waste and emissions
✓ Recycling building materials
✓ Supporting the use of local resources
✓ Implementation of quality assurance and environmental management systems

In The Netherlands, since 1999, it means

✓ Consume a minimum amount of energy and water over the life span
✓ Make efficient use of raw materials
✓ Generate a minimum amount of pollution and waste
✓ Use a minimum amount of land and integrate well with the natural environment
✓ Meet user needs now and in the future
✓ Create a healthy indoor environment

In the United Kingdom, since 2000, it means

✓ Minimising the consumption of carbon-based energy
✓ Improving whole life value by supporting best practice construction
✓ Delivering buildings and structures that provide greater satisfaction, wellbeing and value to customers and users
✓ Respecting and treating stakeholders more fairly
✓ Enhancing and better protecting the natural environment
✓ Being more profitable and using resources more efficiently

Source: Adapted from EU (2001) and HM Government (2008)

In practical terms, sustainable construction can be reduced to three important messages for the way the industry should work.

- Buildings and infrastructure projects should become more cost effective to produce and run, because they have been constructed with less and yield more.
- Construction projects should contribute positively to their environment, using materials and systems that are easily replenished over their full life cycle.
- Contractors and clients should, wherever possible, create higher standards of respect for people and communities involved with the project, from the site workers through to the final community of users.

A central purpose of this text has been to encourage students of construction to engage with modern economic analysis – and, in particular, analysis that includes, at the very least, a study of economic efficiency, environmental externalities and social equity. As Professor Charles Kibert (1999: 3–4) one of the first economists to address the question of sustainable construction perceptively pointed out: 'Economics is of crucial importance in dealing with the subject of sustainability, both to demonstrate the ultimate advantages of creating a sustainable built environment as well as to demonstrate the greatly undervalued role that natural systems play in our economic system'.

Table 15.5 confirms the key ideas that have been raised in this text to assist the construction industry move towards achieving sustainable outcomes. From the list it is evident that the discipline of economics has a lot to offer anyone interested in understanding ways to overcome the barriers to sustainability.

Table 15.5 Factors that contribute to sustainable construction

Part of text	Key ideas
Part A Effective use of resources	✓ Competition and efficient pricing ✓ Achieving productive efficiency ✓ Targeting equity, between present and future generations ✓ Partnering and building long-term relationships ✓ Reducing asymmetric information ✓ Taking advantage of economies of scale ✓ Maximising profit (minimising risk)
Part B Effective protection of the environment	✓ Understanding market failure ✓ Using life cycle analysis ✓ Increasing resource efficiency ✓ Waste minimisation ✓ Environmental valuation of externalities ✓ Distinguishing between private costs and social costs ✓ Taking advantage of opportunities to gain competitive advantages by differentiating products
Part C Economic growth that meets the needs of everyone	✓ Understanding interest rates and interpreting expectations ✓ Forecasting and reading economic signals ✓ Encouraging efficient investment into construction for economic growth ✓ Avoiding supply constraints ✓ Understanding macroeconomic management ✓ Recognising the tension between increasing GDP and the ecological footprint ✓ Delivering buildings and structures that provide greater satisfaction, wellbeing and value to clients and users

> **Key Points 15.1**
>
> ❑ Sustainability embraces three broad themes: (a) the environment, (b) the community and (c) the economy.
>
> ❑ To achieve sustainable outcomes, a more holistic approach needs to be developed.
>
> ❑ In several countries pursuing sustainability, construction has been singled out to require special attention. This is partly due to its significant contributions to the economy, partly due to the negative impacts that it makes to any sustainable development agenda, but mainly because construction is so far behind other industrial sectors.
>
> ❑ Understanding the language and concepts of economics is important to anyone who is serious about dealing with the subject of sustainability.

THE ROLE OF MARKETS

Governments across Europe currently tend to use the market to allocate resources in preference to direct government intervention. (We noted this trend in Chapter 2 and reviewed some of the associated policies throughout Part A.) This means that, increasingly, the difficult questions of resource allocation are left to the private actions of individuals and companies, as they seek to maximise respectively their satisfaction and profit. This is certainly the case in the UK today, where a significant amount of trust is placed in the virtue of the market to allocate resources efficiently. However, there is a strong argument to suggest that construction markets are inefficient given the multitude of firms involved in the sector, the local nature of their activity, and the lengthy supply chain involved in the construction process. Indeed, it is well documented that the construction industry tends to be short term in outlook, and slow to adopt innovation and adapt practice through experience. Sustainable construction, on the other hand, is a long-term goal that requires change.

Loose and Tight Couplings

In the specialised literature of construction economics and organisational theory, the industry is frequently defined as a **loosely coupled system**.

> Loosely coupled activities have few variables in common as the units involved in the production process function relatively independently of each other.

In other words, in loosely coupled systems the left hand frequently does not know what the right hand is doing. A good example of loosely coupled activities is the contrast between the manufacture of building materials and the way these materials are actually used when on site. In fact, one of the hurdles to overcome if sustainable construction is to become a reality involves closing the gap between traditional, project-based approaches to more integrated approaches to building involving on-site installation of purpose-made components. The concept of loosely coupled systems builds on the problems of imperfect information that were cited as forms of market failure in Chapters 6 and 10.

A construction project can be thought of as a network of firms that are drawn together to complete a specific set of operations on site. As the construction process is mainly about co-ordinating specialised tasks, loose couplings can occur across a number of relationships: among individuals, between subcontractors, at the organisational level, and between developers and users. The problems generated by these loose couplings is compounded in the construction industry as the activity is inevitably carried out at a specific and localised site where different teams have been brought together to complete a unique project. This means that in most cases the working relationships on site are based largely on trust in each others judgement; and if a project fails, the trust relationship can quickly sour and turn into an adversarial relationship based on blame. In other words, each project involves a temporary network of firms coming together for the sole purposes of the specific project. At the same time the firms are possibly involved in other projects in which they are expected to co-ordinate their activities and resources with a completely different set of other firms. For example, in Figure 15.3 firm C needs to consider at least three couplings.

Figure 15.3 A network of construction projects

Source: Adapted from Dubois and Gadde (2002: 625)

1 It needs to co-ordinate with other firms involved in firm C's individual projects – represented by the projects employing resources C1, C2, C3 and C4.
2 It needs to co-ordinate with the firms involved in its supply chain, for example firms D and E.
3 Individual projects exist within a broader, more loosely connected, permanent network and firm C needs to co-ordinate with associated firms A and B on activities and resources that lie beyond the scope of each individual project.

In contrast to the construction sector, manufacturing industry tends to be typified by **tightly coupled systems**. Manufacturing usually involves fewer independent elements and a far greater level of co-ordination from a centralised management team. This in turn encourages the use of standardised procedures, uniform product quality, improved management and economies of scale. Long-term relationships, formed in doing the same task with the same team, also encourage product development and learning. David Gann (2000) argues that it is the difficulties of creating the tight couplings enjoyed by manufacturing that lie at the heart of the problems faced by construction, where a lack of integration and innovation are serious constraints.

To achieve a shared vision, clear goals and an industry-wide strategy requires a good level of communication between the participating organisations. This explains why such a great deal of promise is held in the new forms of partnership that were discussed in Chapter 6 (see pages 88–92). Partnering, whether formalised through arrangements such as PFI or simply achieved by informal relationships, can potentially improve the integrated view. For example, the Dutch, Swedes and Danes have bridged many of the gaps between organisations in the house building sector. It is becoming the common practice in these countries to form multidisciplinary teams early in a project to brainstorm how they may contribute to the sustainable construction process. The strength of this type of network is the team building that emerges across disciplines, building links between developers, architects, contractors and, even, tenants. This tightening of the relationship between firms has enabled the Scandinavian and Dutch countries to progress furthest in the transition towards sustainable construction. Indeed they claim to be some of the few countries in the world to achieve economic growth over the past 20 years without increasing the consumption of fossil energy (Jensen and Gram-Hanssen 2008).

In the UK and elsewhere there is still a mindset which tends to fragment the responsibility for sustainable construction. Construction firms argue that they can only adopt holistic approaches if clients ask for them, developers imply that there is no demand for sustainability and investors are hesitant to fund risky new ventures. There is a so-called 'circle of blame' that perpetuates the existing traditional approach to construction. The basic idea is captured in Figure 15.4, which characterises four typical views from the industry. This figure highlights the conservative nature of the construction industry, and serves as a good reminder of the argument introduced in Chapter 9 where we suggested that the firms who choose to adopt the product differentiation message will be the first to break out of this vicious circle.

Figure 15.4 The circle of blame

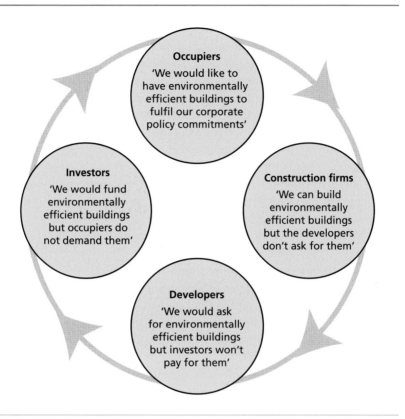

Source: Adapted from Osborne (2001: 84)

Key Points 15.2

❏ Across Europe there is an increasing reliance on market forces to resolve questions of resource allocation in most industrial sectors.

❏ The construction industry is characterised by a low level of inter-firm connection, which means that the same team seldom works together on more than one project.

❏ In the UK and elsewhere there is a tendency to fragment the responsibility for sustainable construction.

❏ The tightening of the relationships between firms is essential if new areas of competence are to emerge in the field of sustainable construction.

THE ROLE OF GOVERNMENTS

According to Manseau and Seaden (2001: 393), in their international review of public policies in Europe, North and South America, South Africa and Japan, government reactions to construction vary from indifference or 'benign neglect' to in-depth reviews. This seems odd, as from the sheer size of the construction industry one might expect that all governments would show some interest in both the sector's development and its performance. In countries pursuing sustainable construction policies, however, there is active government support. Of the fifteen countries covered by Manseau and Seaden's review, eight had specific policies supporting sustainable construction, encouraging lifecycle analysis and/or simply protecting the environment.

In the United Kingdom, successive governments have commissioned a steady stream of reports focusing attention on the construction sector. Indeed, some commentators have even suggested that the industry is suffering from 'initiative overload'. We have referred to five of these reports throughout this book: Latham (1994), Egan (1998), DETR (2000), Fairclough (2002) and HM Government (2008). All five promote ideas to support a more sustainable industry and the three published since 2000 explicitly refer to the urgency required in securing sustainable construction. These reports seemingly find it easy to identify the barriers that need to be removed for an efficient and sustainable industry to emerge. Latham's report is all about building a team between clients, owners, developers, investors, authorities, educators, contractors and users. It is an attempt to discourage participants in the construction process from seeing different purposes and justifications and to counter the strongly ingrained adversarial culture. Published four years later, the Egan Report noted the difficulties posed by the fragmented nature of the construction industry (1998: 14). It argues that this inhibits the opportunities for efficient working, leads the industry to underachieve and limits the possibility of change. The report stressed the advantages of partnering, and the need to eliminate waste and to investigate the ideas of 'lean thinking'. In many ways, Egan paved the way for the first sustainable construction agenda that was launched two years later by the Department of the Environment, Transport and the Regions (DETR 2000). As the government observed in its introduction to the strategy 'the benefits which can flow from a more efficient and sustainable construction industry are potentially immense' (DETR 2000: 7); but unfortunately the industry did not engage wholeheartedly with the ethos or step change involved.

Eight years later the policy was relaunched with a higher profile. The revised strategy was endorsed by industry and government. Interestingly, the 2008 edition was signed by six government departments and supported by a number of public sector clients such as the Highways Agency, Defence Estates and the Housing Corporation. The structure of the strategy involved the setting of specific actions and targets to be met by industry during the next decade. In short, the UK strategy is now expressed as relating to the 'ends' and 'means' of sustainable construction. The 'ends' relate directly to sustainability targets, such as climate change, waste reduction and biodiversity; the 'means' describe processes such as innovation, procurement, regulation and design (HM Government 2008).

In all these reports the government's role as major client is acknowledged. The Fairclough Report (2002: 19–29) went as far as identifying four roles for the government to play:

- as a regulator with regard to aspects such as building and planning regulations
- as a policy maker for issues that directly affect but go wider than the industry, such as energy efficiency, waste management and climate change
- as a sponsor to support research and development and articulate a vision for the future
- and, last but by no means least, as a major client.

It is in its role as a client that the government can make the greatest progress in implementing the sustainability agenda. As Fairclough (2002: 35) pointed out, governments have responsibility for a large variety of public buildings and infrastructure, such as schools and colleges, hospitals and health centres, military installations, prisons, courts, roads, plus specialised buildings such as offices and accommodation.

Even with the development of the private finance initiative (described in Chapter 6) a significant level of government commitment is inevitable. As we explained in Chapter 10, it is due to the nature of public goods that there will always be a significant element of public sector procurement. The point to remember, therefore, is that although private sector firms generate PFI output, it will always be the public sector clients, based in government offices, that initiate these projects. Finally, it is worth remembering that construction has a unique role to play in a sustainable economy because buildings and infrastructure – the products of the industry – have such long life spans. Developments that are under construction today will still be here in 50 – or even 100 – years time.

Complexity

If you have studied the previous chapters – in particular Chapters 1, 6, 7, 8 and 9 – you should recognise that construction is a complex industry compared to other industrial sectors. In these chapters, we demonstrated how construction involves the assembly of a range of materials from a broadly spread range of sources. In turn, each of the material inputs are subjected to different types of processing according to their particular use on a specific project. Assembling these materials are teams of subcontractors gathered for the sole purpose of completing the project. In Chapter 6 we compared the construction process to film production to clarify why each product is regarded as 'unique' and to stress the short-term relationship of the project team. We represented some of the problems associated with this method of production in Figure 15.3. To illustrate the complexity of the construction industry – and the way it shies from the use of standardised components – Egan (1998: 30) referred to the rather startling fact that while the average car contains about 3,000 components, the standard house has about 40,000 parts.

The debate about the practical problems inherent in securing any changes to the construction process has a long history. The industry has become notoriously defensive about the difficulties that emerge from new initiatives and in reviews of the progress of sustainable construction it would not be surprising to see defensive remarks couched in terms of the 'complexities of sustainability'. Research carried out over the last 50 years confirms the complex nature of the construction process – see, for example, Cox and Goodman (1956), Turin (1966, 1975), Winch (1987), Gidado (1996) and Pearce (2003). This research simply underlines the fact that the challenge of engaging the industry in a common vision of any nature is truly difficult.

As emphasised at several points in the text, it is the diverse and fragmented nature of the industry that constrains the opportunities to improve performance, learn from experience and introduce innovation. There are surprisingly few positive references citing opportunities to optimise inputs to produce desired outputs. The construction industry is typified by a simple compartmentalised approach, with each of the specialist trades and numerous gangs of subcontractors tending to be primarily concerned with their own domains. This approach leads to an emphasis on the costs of each component part or process without due consideration to the quality, efficiency and general value for money across the project as a whole. As pointed out in Chapter 1, when we reviewed the ongoing debate between the broad and narrow definitions of construction, it is difficult to know where precisely the boundaries of the industry actually lie. This is why government initiatives to enhance productivity, performance and generally promote sustainability are so important, as the sector is not only complex, but at present it also lacks vision.

Key Points 15.3

❏ There are four roles that government plays in supporting construction: as a regulator, policy maker, sponsor and major client.

❏ The construction process is complex, and this is exacerbated by the diverse, fragmented and conservative nature of the construction industry.

CONCLUDING REMARKS

As an industry, construction clearly needs a cultural change and this has been recognised for a many years. Indeed the European Union in its first broad agenda for sustainable construction recommended that 'employers and professional bodies should phase out recognition of training and educational courses which do not take adequate account of the topic of sustainable construction' (EU 2001: 23). Certainly training, research, collaboration and working as a team could be major drivers of change. Arriving at a consensus on what should be delivered under the heading of construction economics at universities should also help to close the gap between government rhetoric and construction industry practice.

The new approach outlined in this book goes someway towards meeting this requirement. It has introduced some of the strategies that could be adopted by professionals entering traditional construction firms. We suggest four options following *The Greening of Industry for a Sustainable Future*, a framework produced by Schot et al. (1997). These are:

- the defensive option
- the offensive option
- the eco-efficiency option
- the sustainability option.

These four options offer a useful framework for analysing whether sustainable construction can become a reality.

THE DEFENSIVE OPTION

This option refers to output being changed to comply with regulations, on the basis that a primary motivation to make improvements is to avoid costs that might be imposed by non-compliance. In this context, Stern (2007: 434) referred to an OECD study that confirmed that 19 out of 20 countries surveyed had mandatory building regulations (codes) designed to increase energy efficiency. This has been an important strategy in reducing the environmental impacts of business activity. At the very least, it has caused most construction firms to acknowledge some of the environmental aspects of the agenda, and it has had some impact on the use of landfill and virgin aggregates which are both subject to tax. From a client's perspective, it is the prospect of potentially more onerous legislation in the future that is leading some developers and investors to adopt a 'beyond compliance' culture, either to achieve higher returns on projects or to reduce risks. In short, as Sayce et al. (2007: 630) have argued legislation and regulations will be used as important drivers to break into the circle of blame and slowly move us towards greener construction.

THE OFFENSIVE OPTION

This option was introduced in Chapter 9. A firm adopting this strategy would seek to break with convention in order to gain a competitive advantage by differentiating its products from those of its competitors. A few of the major construction firms around the world involved with developing eco-towns such as Curitiba in Brazil, Gaviotas in Colombia, Hammarby in Sweden and BedZED in the United Kingdom have taken advantage of this option and shown that major improvements in sustainability are technically feasible. The exceptional firms involved in these developments have made substantial contributions to their own financial performance while demonstrating the ability to incorporate zero carbon systems into all building types, as well as to creating sustainable approaches to the management of waste and infrastructure. Unfortunately, however, the work of most construction firms falls far short of these standards, and high-profile developments and initiatives to secure sustainable outcomes continue to be exceptions to the rule almost everywhere.

THE ECO-EFFICIENCY OPTION

This strategy is about identifying win-win solutions by reducing environmental impacts and costs. It builds on the ideas presented in Parts A and B of the text and, in terms of construction, it forms the most promising option. It is largely about getting more from less, since it emphasises the importance of achieving value for money as a holistic and long-term objective. These ideas were introduced in the section on value management in Chapter 6 and extended further under the heading of resource efficiency in Chapter 9, where we gave a range of examples of how to increase efficiency in buildings and infrastructure by a factor of four. For instance, Italy and the Czech Republic both continue with notoriously low recycling rates, which are due, in part, to a strong consumer preference for virgin aggregate materials and the lack of any significant price difference between virgin and recycled materials. In contrast, construction firms in the UK and Sweden face higher aggregate taxes which encourage far better rates of recycling (EEA 2008: 8). The UK and Sweden could therefore claim to be relatively more eco-efficient. In other words, enlightened companies in the right cultural setting understand this option, but the vast majority of small firms remain indifferent and uninvolved.

THE SUSTAINABILITY OPTION

This is the most difficult option as it seeks to achieve triple bottom line benefits – addressing economic viability, environmental soundness and social responsibility. Any business in any sector would find this difficult, as most economic activity still takes place with very little concern for the wider environmental and social effects. This option demands adopting new values – values that are more holistic, that show concern for future generations and which delegate responsibility to the lowest level. Perhaps surprisingly, there are some developments taking place within construction that suggest that some of the players in the industry could meet these criteria. For example, Carillion plc used its PFI hospitals as case studies to demonstrate that environmental and social benefits can be explicitly shown in the company accounts alongside the traditional business costs and benefits (Forum for the Future 2002). Reading 4 also implied some strategic developments in corporate social reporting to suggest that some of the larger construction companies are beginning to engage with widening the set of criteria on which they promote and support their business.

The Future

The sustainable construction agenda has made, and is continuing to make, the industry think about a broader range of strategies. It is possible to see players within the construction industry take up opportunities within each of the four options listed above. Clearly far more needs to be done to integrate inputs and outputs in a holistic and schematic manner. It is important, therefore, that at some point in the near future, all construction firms, and not just the brave few seeking to gain a competitive advantage, heed the government messages. At least students of this new approach should recognise the value of breaking the mould and adopting values that could carry this forward. As the literature on sustainability makes clear, it is those working at the present time who determine the quality of life of future generations.

Reading 6

In 2007 the international journal *Construction Management and Economics* celebrated 25 years of publication. In this time it has published approximately 1,200 papers including notes and comments (Pietroforte et al. 2007). Less than 5 per cent of these were related to the sustainable construction debate. This lack of interest is not so much a reflection of the journal but of the lack of activity amongst the established academic research community that submits papers for publication and, more worryingly, the inertia that typifies the construction sector. So an important objective of this final reading is to encourage those completing a study of this text to engage with the new approach and identify ways to change the industry and make it contribute more to sustainable development. The final extract will provide some stimulus as it points to several barriers that explain why sustainable construction is not yet standard industry practice (although other issues exist that the authors overlook). The closing aim, therefore, is to encourage you to think broadly about the nature of the problems that need to be resolved before the journey towards sustainable construction can progress.

Mohamed Matar, Maged Georgy and Moheeb Ibrahim (2008) 'Sustainable construction management: introduction of the operational context space', *Construction Management and Economics* 26: 261–75

Introduction

As a discipline, sustainable construction has been evolving since the late 1980s. It continuously gains momentum as increasing evidence about depletion of the environment and environmental loadings becomes obvious. However, regardless of its importance and the expanding foundation of knowledge in the field, sustainable construction is by no means standard industry practice in many world countries (Landman, 1999). Although environmental protective measures, environmental management systems (EMS) and frameworks have become very common at manufacturing and industrial facilities, only a few construction companies have considered the use of a full EMS system in their projects (Christini et al 2004). Several barriers contribute to hindering sustainable construction being the dominant trend of the industry. Principally, these barriers can be summarized into two main categories: general barriers and technical barriers (Matar et al 2004; Matar 2007).

General barriers to sustainable construction

(1) The lack of expressed interest from different project parties. A survey made in the US by the Reed Research Group revealed that almost 60% of industry professionals do not even try to attempt green projects. Only 32% of construction clients have shown interest in pursuing green construction (RRG 2003).

(2) The lack of training/education in sustainable design/construction (Du Plessis 2002).

(3) The slow recovery of investment in technology to promote sustainable construction practices, such as off-site manufacture (Du Plessis 2002; RRG 2003).

(4) The higher initial cost of sustainable building (Landman,1999).

Technical barriers to sustainable construction

(1) The lack of a well-defined set of sustainable construction practices that can be practically engineered into projects (Pearce and Vanegas 2002). Although several best practice guidelines exist covering different environmental performance aspects, there is

little formally structured information about procedures associated with the inclusion of environmental issues in the construction process (Cole 2000).

(2) The need for a mature and integrated framework of application for sustainable practices in construction. While certain efforts have been ongoing to control and enhance individual aspects of the environmental qualities of built facilities, detailed comprehensive approaches have generally been lacking. Each of the currently available systems has its own set of assumptions and limitations, and is designed for utilization by specific participants in the building process (Scheuer and Keoleian 2002). This hinders the capacity of different industry stakeholders to cooperate in a uniform and positive way.

(3) The disagreement about an optimum project delivery structure......Very few studies have considered the effect of project planning on the capacity to adopt sustainability practices (Chen et al, 2005). As sustainability is concerned with the complete life cycle, a building needs to be considered from its design and construction, through the operational stage, to its deconstruction......

(4) The need for effective drivers for change for different parties in the construction industry (Vanegas and Pearce, 2000; Du Plessis, 2002).

References

Chen, Z., Li, H. and Wong, C.T.C. (2005) Environal Planning: analytic network process model for environmentally conscious construction planning. *Construction Engineering & Management*, 131: 92–101.

Christini, G., Fetsko, M. and Hendrickson, C. (2004) Environmental management systems and ISO 14001 certification for construction firms. *Construction Engineering and Management*, 130: 330–6.

Cole, R.J. (2000) Building environmental assessment methods: assessing construction practices. *Construction Management and Economics*, 18: 949–57.

Du Plessis, C. (2002) *Agenda for Sustainable Construction in Developing Countries: A discussion document*. Report for CIB and UNEP

Landman, M. (1999) Breaking through the barriers to sustainable building: insights from professionals on government initiatives to promote environmentally sound practices, MA thesis, Tufts University, USA.

Matar, M.M. (2007) An integration framework for sustainable construction, MSc thesis, Cairo University, Egypt.

Matar, M.M., Georgy, M.E. and Ibrahim, M.E. (2004) Towards a more applicable set of sustainable construction practices, in *International Conference: Future Vision and Challenges for Urban Development*, HBRC, Cairo, Egypt, 20–22 December pp. CE16: 1–12.

Pearce, A.R. and Vanegas, J.A. (2002) Defining sustainability for built environment systems: an operational framework. *International Journal of Environmental Technology and Management*, 2(1–3), 94–113.

RRG (Reed Research Group) (2003) Where our readers stand on sustainability. *Building Design and Construction*, 11: 14–17.

Scheuer, C.W. and Keoleian, G.A. (2002) *Evaluation of LEED Using Life Cycle Assessment Methods*, National Institute of Standards and Technology (NIST), Gaithersburg, MD, USA.

Vanegas, J.A. and Pearce, A.R. (2000) Drivers for change: an organizational perspective on sustainable construction, in Walsh, K. D. (ed) *ASCE Construction Congress VI*, Orlando, Florida, USA, 20–22 February, pp. 406–15.

....

Extract information: Adapted and edited from pages 261–2 plus relevant references from pages 274–5.

Glossary

adaptive-expectations hypothesis A theory of economic behaviour which states that people's expectations of the future rate of inflation are informed primarily by the rate of inflation in the immediate past. This process is also known as extrapolative expectations, and can be applied to many economic variables.

aggregate demand (AD) All expenditures for the entire economy summed together.

aggregate supply (AS) All production for the entire economy summed together.

allocative efficiency The use of resources that generate the highest possible value of output as determined in the market economy by consumers. Also referred to as economic efficiency.

asymmetric information A situation where two parties to a transaction involving a good or service have unequal knowledge of the properties or risks involved in making that transaction.

average fixed costs (AFC) Total fixed costs divided by the number of units produced.

average total costs (ATC) Total costs divided by the number of units produced.

average variable costs (AVC) Total variable costs divided by the number of units produced.

balance of payments A summary of monetary transactions with overseas nations. It is compiled as an account of inflows and outflows recording visible and invisible trade, investment earnings, transfers and financial assets.

barriers to entry Conditions in the marketplace that make it either impossible or very difficult for firms to enter an industry and offer competition to existing producers or suppliers. Factors which can make it very difficult for a firm to enter a market include the high start-up costs facing companies new to an industry, the legal requirements imposed by government and technological constraints.

base year The year which is chosen as the point of reference for comparison to other years.

base rate The rate of interest that UK financial intermediaries use as a reference for all other interest rates for lending and receiving deposits. For example, large financial intermediaries will borrow from one another at interest rates close to the base rate.

boom A period of time during which overall business activity is rising at a more rapid rate than its long-term trend.

Building Cost and Information Service (BCIS) Set up by the Royal Institution of Chartered Surveyors in 1962, this service provides detailed information on construction costs. The BCIS publishes a building cost index and a tender price index. The data is revised on a quarterly basis. Copies are available from most academic libraries.

building cycles Refers to the fluctuations in construction output – from boom to bust and back again. This sequence has been studied by many economists. The cycles have been measured as varying in duration from 2–25 years according to the sector and country. For example, Ball and Wood (1994) identified housing cycles of 13 years in Germany and 25 years in Sweden.

building regulations A code of practice which specifies the type and minimum quality of materials to be used in a building. These regulations are legally enforced by district councils through, for example, building controls officers.

Building Research Establishment (BRE) A set of government laboratories based in Garston, Watford. The BRE researches general developments, such as those concerning the control of fire, energy and the environment, that relate to all types of construction.

business fluctuations A type of cycle found in overall economic activity, as evidenced by changes in national income, employment and prices. Sometimes referred to as a business cycle.

capital All manufactured resources, including buildings, equipment, machines and improvements to land.

capital goods These are goods that are used in the production of other goods. Examples include cranes, factories and foundries. Consumers do not directly consume capital goods.

capital value The monetary worth of an asset; for example, the price it could be purchased for.

cartel Any arrangement made by a number of independent producers to co-ordinate their buying or selling decisions. The members of a cartel agree in effect to operate as if they were a monopoly.

central bank The official institution of a country which guarantees the liquidity of that nation's banking system. In some cases, the central bank monitors and supervises commercial banks on the government's behalf. It normally acts as banker to the banks and other nationally important institutions. It is usually owned by the government and manages the national debt, exchange rates and the issuing of currency.

centrally planned model A theoretical system in which the government controls the factors of production and makes all decisions about their use and about the distribution of income. This system is often associated with communist nations. A centrally planned economy is also known as a pure command system.

ceteris paribus The assumption that all other things are held equal, or constant, except those under study.

chaos theory This suggests that the natural order of things is stormy and erratic rather than efficient and predictable.

circular flow model A model of the flows of resources, goods and services, as well as money, receipts and payments for goods and services in the economy.

claimant unemployment This is a record of the number of people claiming unemployment related benefits on one particular day each month.

cobweb theorem A dynamic model which tries to explain why cyclical fluctuations in output and prices – such as those in the property sector – can occur.

coincident indicators Economic statistics that are used by economic forecasters to track movements in the economy. For example, changes in output and the stock levels of raw material confirm that an economy is changing.

collusion An agreement, written or unwritten, between producers to determine prices, share out markets and/or set production levels, to avoid the danger of price wars and excessive competition.

commercial banks These are privately owned profit-seeking institutions, sometimes referred to as joint-stock banks to highlight the fact that they have shareholders. Most high street banks, such as HSBC, NatWest and Barclays, are commercial banks.

community infrastructure levy A proposal in the 2008 Planning Bill that would encourage local authorities to charge developers a tariff to contribute towards infrastructure requirements.

competition policy This is a collection of government measures to monitor and control firms attempting to operate in an anti-competitive manner. For example, the government attempts to prevent mergers between companies that are not in the public's interest.

competitive tendering The process of inviting contractors to bid for work on a particular project. The opportunity to tender may be open to all companies or restricted to a set of selected (preferred) contractors.

complementary goods Two goods are considered complementary if both are used together. The more you buy of one, the more you buy of the other – and vice versa. For example, bricks and cement are complementary goods.

Confederation of British Industry (CBI) Founded in 1965, the CBI represents the interests of British firms. Membership consists of thousands of companies, plus hundreds of trade associations and employers' federations. The CBI's main aim is to express business views to government.

concentration ratio A measure of the degree to which an industry is dominated by its largest firms.

constant capital approach This approach to sustainable development argues that man-made capital can act as a substitute for natural resources (natural capital) and, therefore, that as long as aggregate capital stock remains constant, there will be a future.

constant prices Monetary value expressed in terms of real purchasing power, using a particular year as the base, or standard of comparison, to allow for price changes. For example, by expressing GDP at constant prices, comparisons can be made over a number of years.

constant returns to scale A situation in which the long-run average cost curve of a firm remains horizontal as output increases.

consumer goods Goods that are used directly by households to generate satisfaction. Contrast with capital goods.

consumer price index (CPI) A weighted average of prices of a representative set of goods and services purchased by the typical household. It has been the official measure used by the UK government to monitor inflation since January 2004. The CPI is similar to the retail prices index (RPI), but there are differences of coverage and methodology.

consumer sovereignty The concept that the consumer is king. In other words, the idea that consumers ultimately determine which goods and services are produced in the economy. This may not apply in markets dominated by very large firms.

contestable markets This refers to markets in which there is strong potential (or actual) competition because there are no barriers to entry or exit. In a contestable market, firms have to price products competitively and profits are constrained.

contingent valuation method This is a technique used to identify the price of an externality. It involves a survey of the interested parties in an attempt to attribute a hypothetical (monetary) value to an environmental gain or loss.

contracting out Used in the context of privatisation, contracting out refers to the transfer of publicly provided activities to private contractors. For example, a county council may contract out the management of its property portfolio to a private sector surveying firm.

cost-benefit analysis This is a way of appraising an investment proposal. It involves taking into account the external costs and benefits of a proposed development as well as the conventional private costs and benefits. This is done by estimating monetary values for aspects such as health, time, leisure and pollution.

cost-push inflation A rise in price level associated with a rise in production costs, such as the price of raw materials.

cover pricing This describes a situation where a bidder agrees (colludes) with a competitor when placing a tender for work to distort the market by quoting a price that is intended to be too high to win the contract. As a result the tendering authority, such as a local council or other client, is left with a false impression of the level of competition and this may result in it paying a higher price than is necessary.

credit crunch A situation that emerged in the financial markets in 2007 in which lenders began to raise the cost of borrowing and restrict the supply of loans.

current prices Monetary values expressed in terms of today's prices. In other words, what you have to pay for a good or service today. Also called absolute or nominal prices. Contrast with constant prices.

cyclical indicators Economic statistics that are used by economic forecasters to analyse the state of the economy. See entries for leading, lagging and coincident indicators.

deflation This is a sustained persistent fall in the general price level. It is the opposite of inflation.

demand function A symbolised representation of the relationship between the quantity demanded of a good and its various determinants. It looks like an algebraic equation but it is actually just shorthand notation.

demand management Government policies designed to control the level of total demand in an economy. Demand management policies are closely associated with Keynesian economics.

demand-pull inflation An increase in price level caused by total demand exceeding the current level of supply. Demand-pull inflation is often an unwanted outcome of Keynesian policies introduced to achieve full employment.

demand schedule A set of pairs of numbers showing various possible prices and the quantities demanded at each price. The demand schedule shows the rate of planned purchase per time period at different prices of the good.

demand-side economics This term generally refers to government policies which attempt to alter the level of aggregate demand.

depression A period of prolonged and deep recession in the economy, characterised by significant falls in output and soaring unemployment.

deregulation Describes the process by which state monopolies are opened up to competition – a strategy adopted in the UK since 1979. Now used to describe a situation in any industry in which statutory barriers to competition are liberalised or removed.

derived demand Demands created in a market to help meet other demands. For example, the demand for factory buildings is derived from the demand for manufactured goods. The term highlights the distinction between investment goods and consumer goods.

design and build An all-embracing agreement in which a contractor agrees to undertake building, engineering work, design and cost estimating as part of a package for a client.

direct policy This phrase is used to distinguish direct government intervention from broader macroeconomic policies. Direct policy tends to be of a legislative nature.

direct relationship A relationship between two variables that is positive, such that an increase in one is associated with an increase in the other, and a decrease in one is associated with a decrease in the other.

discounting A mathematical procedure by which the value of a sum or a stream of sums due to be received at specific dates in the future is expressed in terms of its current value.

discount rate The interest rate at which the Federal Reserve lends on a short-term basis to the US banking sector.

diseconomies of scale When increases in output lead to increases in long-run average costs.

ecological footprint An accounting tool used to measure how much productive land and water an individual, city or country requires to produce all the resources it consumes and to absorb all the waste it generates.

economic good Any product or service that is scarce and allocated by the price mechanism. Contrast with free good.

economic growth An increase in an economy's real level of output over time. Usually measured by the rate of change of national income from one year to the next.

economic system The institutional means through which resources are used to satisfy human wants.

economics A social science studying human behaviour and, in particular, the way in which individuals and societies choose among the alternative uses of scarce resources to satisfy wants.

economies of scale When increases in output lead to decreases in long-run average costs.

effective demand Demand that involves desire and ability to pay. In other words, it is the demand that can be measured by actual spending.

efficiency See glossary entries for allocative efficiency and productive efficiency.

elasticity A measure of the level of response to a specific change in market conditions such as a change in price or income.

embodied energy The energy used in the process of extracting raw materials, manufacturing components and transporting them to site.

endogenous variables These are economic factors which affect other aspects of a theory or model from within. For example, the level of unemployment will affect the amount of income tax collected. Contrast with exogenous variables.

entrepreneur The fourth factor of production involving human resources that perform the functions of raising capital, organising, managing, assembling the other factors of production (land, capital and labour), and making basic business policy decisions. The entrepreneur is a risk-taker.

environment In modern economic analysis, environmental assets such as clean air, species diversity and tropical rainforests are considered alongside the allocation of resources of traditional economic goods and services. Also see the entry for environmental policy.

environmental economics The origins of environmental economics date back to the 1960s when green thinking became a popular concern. A central tenet of the environmental economics school is that as the economic system cannot operate without the support of the environment, financial value should be placed on environmental services.

environmental policy Governments across the world are accepting responsibility for the global and local problems that arise as a result of market failure. In the UK, for example, the government has set out a strategy for sustainable development in the general economy (DETR 1999) as well as a sustainable strategy for the construction industry (DETR 2000).

equilibrium A situation in which the plans of buyers and sellers exactly coincide, so that there is neither excess supply nor excess demand.

equilibrium price The price that clears the market, at which there is no excess quantity demanded or supplied. In other words, the price at which the demand curve intersects the supply curve. Also known as the market clearing price.

equity See glossary entries for horizontal equity and vertical equity.

exogenous variables These are economic factors which impinge upon a theory or model from the outside, such as the weather. They are sometimes referred to as autonomous variables, and they contrast with endogenous variables.

expenditure approach A way of computing national income (GDP) by adding up the values of all spending on final goods and services.

external economies of scale These are the savings a firm can achieve in long-run average costs due to changes that benefit the *whole* industry. For example, a big increase in state-funded training could help all firms in a particular sector become more efficient.

external costs The costs experienced by third parties due to the production or consumption of a good or service; not usually expressed in monetary terms.

externalities The benefits or costs that are experienced by parties other than the immediate seller and buyer in a transaction. These third-party effects are also known as external costs or benefits.

factor markets In factor markets, households sell resources such as labour, land, capital and entrepreneurial ability. Businesses are the buyers of these resources to generate output (see Figure 3.1).

factors of production Often grouped under four headings – land, capital, labour and entrepreneurial ability – these are the resources and inputs required to produce any good or service.

financial intermediaries Institutions such as commercial banks and building societies that link savers (people and organisations that have extra funds) with borrowers (people and organisations that need funds).

Financial Services Authority The organisation responsible with the Bank of England for regulating and supervising all financial intermediaries in the United Kingdom.

fine tune A term used in economics to suggest that an economy has an engine like a car, and this can be set up to run at different levels of efficiency by adjusting various flows.

fiscal policy A combination of government spending and taxation used to achieve macroeconomic management.

fixed costs The costs that do not vary with output. Examples of fixed costs include the rent on a building and the price of machinery. These costs are fixed for a particular period of time, although in the long run they are variable.

foreign exchange market The market for buying and selling foreign currencies.

free enterprise A system in which private businesses are able to obtain resources, to organise those resources and to sell the finished product in any way they choose.

free good Exceptional goods, such as watercourses, sunshine and air, that have a zero price because they do not require the use of scarce factors of production.

free market model A theoretical economic system in which individuals privately own productive resources and can use these resources in whatever manner they choose. Other terms for this system are a pure market or pure capitalist economy.

free riders Individuals who do not pay for the goods and services that they consume.

full employment This represents a situation where the number of job vacancies exceed the number of people registered as unemployed. As a concept it recognises the fact that even when the overall economy is doing well there will be some workers between jobs searching for vacancies that suit their skills.

game theory The study of how decisions are made, particularly in the context in which decision-makers have to take other people's responses into account. It is used to analyse oligopolistic forms of competition.

golden rule A fiscal objective that constrains a government only to borrow to finance investment not to fund current spending.

government failure The concept that government policy intervention may not necessarily improve economic efficiency.

government intervention Measures undertaken by the state to achieve goals not guaranteed by the market system.

great depression The worldwide economic downturn that started in the United States in 1929 and led to the most severe depression in the world to date.

gross domestic fixed capital formation A national income accounting category representing expenditure on fixed assets (such as buildings, vehicles, plant and machinery). It is more common to refer to this expenditure as investment. Spending on maintenance and repairs is excluded from this measure.

gross domestic product (GDP) The most common measurement of a nation's wealth, based on income generated from resources *within* its own boundaries; the monetary value of its output of goods and services.

gross national income (GNI) A measurement of a country's wealth. It represents the total output of goods and services produced by the country in a year, plus the net value of overseas assets. In the UK, there is little difference between GDP and GNI. In 2000, UK GDP was £950 billion and GNI was £956 billion.

harmonised index of consumer prices (HICP) A standard measure of consumer price inflation that allows comparison between European Union countries. The series commenced in January 1996 and became the official target of government policy in 2004. It is usually referred to as the consumer price index.

headline inflation rate The change in the retail price index that is announced by the UK government in a monthly press release. It contrasts with the official rate of inflation, which is based on the consumer price index.

hedonic pricing method A technique used to identify the price of an externality. It provides an estimate of the implicit price of an environmental attribute by comparing the value of two identical goods, one with the environmental element and the other without.

horizontal equity The concept that all people should be treated identically. The idea that underpins equal opportunities policies.

human capital Investment in education and training that enhances the productivity of individuals.

ILO unemployment rate A measure of unemployment produced by the International Labour Organisation. It defines unemployment as people who are without work yet actively seeking employment. Data is gathered through labour force surveys.

imperfectly competitive markets A broadly used term to refer to all markets that do not have the characteristics of perfect competition. Commonly referred to as imperfect competition.

income approach A way of measuring gross domestic product by adding up all factor rewards – that is, the sum of wages, interest, rent and profits.

income elasticity of demand A measurement of the responsiveness of the quantity demanded to a change in income.

index numbers A way of expressing the relative change of a variable between one period of time and some other period of time selected as the 'base year'. For example, the base year index number is set at 100 and the value of the variable in subsequent years is expressed above or below 100 according to its percentage deviation from the base.

inferior goods Products or services for which demand decreases as consumer income increases. Goods of this nature are exceptions. Contrast with normal goods.

inflation A sustained rise in prices, formally measured by the consumer price index.

injections Supplementary inputs into the circular flow of income in an economy. Typical injections are investment, government spending and export income.

interest rates These determine the cost of obtaining credit and the rewards paid to owners of capital.

internal economies Formal term for economies of scale. Contrast with external economies of scale.

inverse relationship A relationship between two variables in which an increase in one variable is associated with a decrease in the other, and a decrease in one variable is associated with an increase in the other.

investment Spending by businesses on things like machines and buildings which can be used to produce goods and services in the future.

joint production A term coined to denote the phenomenon that several outputs inevitably emerge from economic activity.

labour The human resource involved in production. In other words, the contributions (both thinking and doing) of people who work.

lagging indicators Economic statistics (such as unemployment and investment) that change approximately twelve months after a change in overall activity (gross domestic product).

land One of the factors of production. In economic terms, land consists of both the physical space (in which economic activity can be located) but also natural resources such as coal, oil, water, natural vegetation and climate.

law of demand Quantity demanded and price are inversely related. In other words, the law of demand states that more is bought at a lower price, less at a higher price (other things being equal). Also known as the theory of demand.

law of diminishing (marginal) returns States that after some point successive increases in a variable factor of production, such as labour, added to fixed factors of production will result in less than a proportional increase in output.

law of increasing opportunity costs An economic principle that states that in order to get additional units of a good, society must sacrifice ever-increasing amounts of other goods. It is also referred to as the law of increasing relative costs.

law of supply The relationship between price and quantity supplied (other things remaining equal) is a direct one. That is, as price increases so does the quantity supplied and vice versa.

leading indicators Economic statistics (such as retail sales and consumer credit) that change approximately six months in advance of gross domestic product and are used to predict changes in the economic cycle.

leakages A withdrawal from the circular flow of income. Examples are net taxes, savings and imports.

liabilities All the legal claims for payment that can be made on an institution or company. In short, the amount owing to others.

life cycle analysis This approach aims to take into account the whole life costs incurred during a project. The analysis covers the total cost of ownership, including the initial investment, and production, operation and maintenance costs.

liquidity This describes the ease with which an asset can be used to meet liabilities. Cash is the most liquid asset.

long run That time period in which all factors of production can be varied.

long-run average cost curve (LAC) This represents the cheapest way to produce various levels of output given existing technology and current prices. It is derived from a compilation of short-run positions (see Figure 7.4b, page 115).

loosely coupled systems Every industrial activity is to some extent interdependent with a number of other activities; that is, the activities are coupled in some way. Some of these couplings are 'tight' while others are 'loose'. Construction is typified as a loosely coupled system as the firms involved in the industry operate fairly independently of another.

low inflation A trend for annual price increases to be below 5 per cent.

macroeconomics The study of economy-wide phenomena, such as total consumer expenditure.

macroeconomic objectives Targets relating to the whole economy, such as price stability, employment levels and the balance of payments.

marginal cost (MC) The change in total costs due to a one-unit increase in the variable input. The cost of using more of a factor of production.

marginal physical product (MPP) The output that the addition of one more worker produces. The marginal physical product of the worker is equal to the change in total output that can be accounted for by hiring one more worker.

marginal propensity to leak (MPL) The proportion of an increase in national income that is withdrawn from the circular flow. For example, a 0.2 MPL indicates that out of an additional £100 earned, £20 will 'leak' in imports, savings or tax.

marginal revenue (MR) The change in total revenues resulting from the sale of an additional unit of a product.

market An abstract concept concerning all the arrangements that individuals have for exchanging goods and services with one another. Economists often study the market for particular goods and service, such as the labour market, the car market, the commercial property market, the housing market, the building materials market, the credit market, and so on.

market-based instruments These involve various incentive systems designed to operate through the price mechanism to encourage environmentally friendly behaviour. Examples include the climate change levy and the landfill tax.

market-clearing price Another term for equilibrium price.

market economy An economy in which prices are used to signal the value of individual resources to firms and households.

market failure A situation in which the free forces of supply and demand lead to either an under- or over-allocation of resources to a specific economic activity.

market mechanism See market economy and price mechanism.

market structures The characteristics of a market which determine the behaviour of participating firms, such as the number of buyers and sellers and the ease of entry into (and exit from) a market.

market supply schedule A set of numbers showing the quantity supplied at various prices by the firms comprising the industry. The horizontal summation at each price suggests the market supply.

materials balance model This model provides a framework to describe the relationships between the economy and the environment by focusing on natural resource depletion and pollution.

maximum price legislation A price ceiling set by a government agency specifying a level in a specific market beyond which prices must not rise.

menu costs The resources used up revising contracts due to inflation.

merit good A product or service that is deemed socially desirable by politicians. If left to the private market, merit goods may be under produced.

mesoeconomics A study of economic activities at the level of sectors or industries.

microeconomics The study of economic behaviour of individual households and firms and how prices of goods and services are determined.

midpoint method A mathematical technique used for estimating elasticity. It involves calculating percentage changes in price and quantity compared with the average of the initial and final values.

minimum efficient scale The lowest rate of output at which long-run average costs reach their minimum point.

mixed economy An economic system in which decisions about how resources are used are made partly by the private sector and partly by the government.

mobility of labour The ease with which labour can be transferred from one type of employment to another. Mobility of labour is considered in terms of geographical and occupational mobility. The converse concept, the immobility of labour, is often employed by economists.

models Simplified representations of the real world used to make predictions and to provide greater clarity to economic explanations.

modern methods of construction The phrase covers a wide range of products and solutions, encompassing everything from new innovative building components to entire factory-built structures and building modules.

monetarists Economists who believe that changes in money supply are important in the determination of the full employment level of national income.

monetary policy A policy, usually implemented by the central bank, to control inflation rates by influencing aggregate demand through changes in interest rates.

monetary policy committee (MPC) A Bank of England committee established in 1997 that sets interest rates independently of HM Treasury in order to achieve the UK government's predetermined target rate of inflation.

money Anything that is generally accepted as a means (medium) of payment for goods and services, or the settlement of debts (that is, deferred payments). Ideally, money should act as a store of value and unit of account, although during periods of high inflation it may become deficient in these respects.

money supply A generic term used to denote the amount of 'money' in circulation. There are many definitions of money supply, with individual variants including different types of bank deposits in the overall money supply measure.

monopolistic competition A market situation in which a large number of firms produce similar but not identical products. There is relatively easy entry into the industry.

monopoly A market structure in which a single supplier dominates the market.

monopsony A market in which there is only one buyer.

multiplier The number by which an initial injection into an economy must be multiplied to find the eventual change in national income. Mathematically, it is the reciprocal of the marginal propensity to leak (MPL).

national accounts An annual record of an economy's aggregate performance.

National Audit Office (NAO) A government agency responsible for auditing the tax revenues spent by central government. It investigates all aspects of value and publishes around 60 reports each year on its findings. Electronic versions of these reports are freely available online for download (see Part C web reviews on page 202 for details,).

national income A generic term for all that is produced, earned and spent in a country during one year. Strictly speaking, it is defined as gross national product (GNP) minus capital depreciation.

nationalisation Taking into public ownership some part, or all, of an economic activity previously located in the private sector.

natural capital approach An approach to sustainable development that regards the environment as critically important to future economic wellbeing. Its basic premise stands in stark contrast to the constant capital approach.

natural monopoly An industry best suited to production through a single firm. Such situations arise when production requires extremely large capital investments.

neoclassical economics Economists in the neoclassical school follow the traditions of classical economists such as Adam Smith and believe that free markets are best suited to steer economies.

neutral equilibrium A theoretical concept closely associated with a two sector economy where the established levels of activity persist forever, since there are no pressures for change.

new economy Coined in the late 1990s, this term describe the use of information technology in business, both within and between firms and between firms and consumers. More generally, it describes a way of restructuring economies through the use of information technology.

nominal values The values of variables such as gross domestic product and investment expressed in current prices. Also called money values. In other words, the actual market prices at which goods and services are sold.

normal goods Goods for which demand increases as income increases. Most products and services that we deal with are normal goods.

normal profit The minimum rate of profit necessary to ensure that a sufficient number of people will be prepared to undertake risks and organise production. In more formal terms, it is the normal rate of return on investment – which differs from industry to industry.

normative economics Analysis involving value judgements about economic policies; relates to whether things are good or bad. A statement of 'what ought to be'.

notes and coins The currency of a nation, normally referred to as cash.

novation A term coined during the early 1990s to describe a client-led contractual bonding between an architect and building contractor. See design and build.

Office for National Statistics (ONS) The UK government agency responsible for compiling and distributing economic, demographic and social statistics. It publishes an array of economic data on the Internet (see page 28 for details).

official rate of inflation In most industrialised countries, the target, or, typically, the midpoint of a target range, for consumer price inflation is between 1 per cent and 2.5 per cent. In the UK, the current official target rate of inflation is 2 per cent as measured by the consumer price index.

off-site production (OSP) A term used to describe any construction process that is carried out away from the building site, such as in a factory or in a specially created temporary production facility close to the construction site.

oligopoly A situation in which a large part of the market is supplied by a small number of firms. The firms may behave as if they are interdependent.

OPEC *see* Organisation of Petroleum Exporting Countries.

opportunity cost The highest-valued alternative that has to be sacrificed for the option that has been chosen.

opportunity cost of capital The normal rate of return or the amount that must be paid to an investor to induce him or her to invest in a business. Economists consider the opportunity cost of capital as a cost of production.

optimism bias A measure of the extent to which actual project costs (and project duration) might exceed the original estimates.

Organisation for Economic Cooperation and Development (OECD) The OECD publishes many economic commentaries each year. It produces data allowing international comparison of economic and environmental indicators. Based in Paris, the OECD's members comprise all the major capitalist countries, including Australia, Japan, the United States, Canada and western European nations.

Organisation of Petroleum Exporting Countries (OPEC) This is a consortium of 13 petroleum producing nations. The organisation was formed in 1960 to control oil supply and prices. It is an example of a cartel.

output approach A way of measuring national income (see gross national product) by adding up the value of the output produced by each sector of the economy.

partnering A broadly defined term to describe a situation in which two or more organisations work openly together to improve performance by agreeing mutual objectives and ways for resolving any disputes. See also public-private partnership and strategic partnering.

per capita A Latin phrase meaning per head of the population.

perfect competition A market structure in which the decisions of buyers and sellers have no effect on market price.

perfectly competitive firm A firm that is such a small part of the total industry picture that it cannot affect the price of the product it sells.

perfectly elastic A supply or demand curve characterised by a reduction in quantity to zero when there is the slightest increase or decrease in price. Producers and consumers are completely responsive to any change of price.

perfectly inelastic The characteristic of a supply or demand curve for which quantity supplied remains constant, no matter what happens to price. Producers and consumers are completely unresponsive to price changes.

planning curve Another name for the long-run average cost curve.

planning gain A trade-off agreement between a local authority and developer that permits the developer to build in return for some compensatory benefits to the community. A common example involves the provision of social housing as an integral part of a proposal for a commercial housing development.

planning horizon Another name for long-run cost curves. All inputs are variable during the planning period.

planning regulations Each local authority has a set of plans on how its area will develop and a committee (or body) responsible for deciding what is allowed to be built and where. The terms of reference are determined by centrally produced planning policy guidance notes.

polluter pays principle A strategy based on market incentives to assure that those who pollute are encouraged to reduce the costs that fall on society. The principle was succinctly set out in 1987 by the Secretary of State for the Environment: 'The polluter must bear the cost of pollution.'

positive economics Analysis that is strictly limited to making purely descriptive statements and scientific predictions, such as 'if A, then B'. Positive statements can be checked against the evidence – they are statements of what is. Contrast with normative economics.

precautionary principle A recommendation that the costs of avoiding unacceptable environmental consequences should be met even if the precautions may turn out in the long run to have been greater than necessary.

present value A future monetary value expressed in today's prices; the most that someone would pay today to receive a certain sum at some point in the future.

price-elastic demand When a price change of a product or service results in a more than proportionate change in demand, then demand is said to be price elastic.

price elasticity A measurement of the responsiveness of the quantity demanded or supplied to a change in unit price.

price elasticity of supply A measurement of the responsiveness of the quantity produced for the market due to a change in price.

price index The cost of today's basket of goods expressed as a percentage of the cost of the same basket during a base year.

price-inelastic demand When a change in price results in a less than proportionate change in demand, then demand is said to be price inelastic.

price mechanism Prices are a signalling system between firms and households concerning the use of resources. If a price mechanism operates, there is a market economy. The terms 'price' and 'market' are interchangeable.

price system An economic system in which (relative) prices are constantly changing to reflect changes in supply and demand for different commodities. The prices of commodities are signals to everyone within the system about which are relatively expensive and which are relatively cheap.

price-taker A characteristic of a perfectly competitive firm. A price-taker is a firm that must take the price of its product from the prices of its competitors.

principal-agent A concept used to highlight one possible cause of market failure – inexperienced clients (the principal) cannot be confident that their best interests are properly represented by contractors (the agent) acting on their behalf.

principle of exclusion This simply means that anyone who does not pay for a product or service will not be allowed to benefit from consuming that particular good – they will be left out or excluded from the product's or service's benefits.

principle of rivalry The principle that private goods cannot be shared. If person A uses a private good, than that prevents the possibility of person B using that good. Persons A and B cannot eat the same apple simultaneously.

private commercial A sector of the construction industry concerned with privately funded commercial developments such as shops, offices and leisure facilities (see Tables 1.3 and 1.4).

private costs The specific monetary payments made by firms and individuals in the production and consumption of goods and services.

private finance initiative (PFI) A modern form of procurement to encourage private investment in public sector projects. PFI suppliers are typically contracted not only to construct a facility but also to manage and maintain the infrastructure and provide support services for a period of several years following construction. Introduced in the UK in 1992, the initiative is designed to reduce the pressure on public sector borrowing.

private goods Goods that can only be consumed by one individual at a time. Private goods are subject to the principles of exclusion.

private industrial A sector of the construction industry concerned with privately funded industrial developments like factories and warehouses. Since the privatisation of the public utilities – gas, water, and electricity supply and generation – the significance of this sector has increased (see Tables 1.3 and 1.4).

privatisation In very general terms, this involves the transfer of assets from the public sector to the private sector.

procurement A generic term used by professionals within the built environment to describe the general process of obtaining, acquiring and securing some property or land.

product differentiation This occurs in imperfect markets when individual producers introduce differences in the characteristics of products and services that are broadly alternatives to each other in order to gain a competitive advantage. In construction markets, product differentiation may be achieved by craftsmanship, foreign language ability, associated financial packages or some technological superiority. A differentiated product gives producers greater freedom to determine their prices.

product markets A market in which businesses sell consumer goods to households (see Figure 13.1).

production function The relationship between inputs and output. A production function is a technological, not an economic, relationship.

production possibility curve A curve representing the maximum combinations of two goods that can be produced assuming (a) a fixed amount of productive resources and (b) efficient use of those resources.

productive efficiency The utilisation of the cheapest production technique for any given output rate; no inputs are wilfully wasted. Also known as 'technical' efficiency.

profit The income generated by selling something for a higher price than was paid for it. In production, profit is the difference between total revenues received from consumers who purchase goods and the total cost of producing those goods.

profit maximisation model A model based on the assumption that the central aim of the firm is maximising profit. This can be achieved by setting prices so that marginal revenue equals marginal cost. If marginal revenue is greater than marginal cost, total profit can be increased; conversely, when marginal cost is greater than marginal revenue, total profits decline.

progressive income tax A tax system in which a higher percentage of income is taxed the more a taxpayer earns. Put formally, the marginal tax rate exceeds the average tax rate as income rises.

project partnering An arrangement between the main contractor and the client working together on a single project. It embraces a range of possibilities, but it is usually comes into effect after the contract for the project has been awarded. It has been devised to overcome the adversarial relationships that typify traditional construction contracts. It is designed to prevent contractors regarding a project as a sequence of separate operations and making no long-term commitment to its success.

public goods Goods for which the principles of exclusion and rivalry do not apply. Public goods can be jointly consumed by many individuals simultaneously, at no additional cost and with no reduction in the quality or quantity of the provision concerned.

public (non residential) A sector of the construction industry concerned with the construction of roads, prisons, schools, etc. In short, public sector works other than housing (see Tables 1.3 and 1.4).

public private partnership This is a particular type of contractual arrangement between the public sector and private sector firms. These give the private sector a greater role in financing, building and maintaining public sector facilities, although the government retains a stake in the PPP company. In contrast to the private finance initiative, under public private partnership arrangements the government is not liable for a fixed stream of annual payments.

public sector The simplest definition is all forms of ownership by central and local government.

public sector net cash requirement (PSNCR) This is a measure of the public sector's short-term financing requirement. It covers the combined funding requirement of both central and local government. The concept was previously known as the public sector borrowing requirement (PSBR).

quasi-public goods Goods or services which by their nature could be made available for purchase by individuals, but which the state finds administratively more convenient to provide for all the nation (such as roads).

rational-expectations hypothesis A theory that suggests that individuals combine all available information to form judgements about the future.

real values Measurement of economic values after adjustments have been made for inflation.

recession A period of time during which the rate of growth of business activity is negative, or consistently less than its long-term trend. If it is unduly prolonged, a recession may become a depression.

refinancing rate The interest rate at which the European Central Bank lends on a short-term basis to the euro banking sector.

registered social landlords A group of private organisations that manage nearly two million homes in the UK for tenants on lower incomes with support from the government. Examples of organisations on the register include charitable companies, housing associations and co-ownership societies.

rent controls A price ceiling on rents charged for private rented accommodation. First introduced in 1915, rent controls were designed to protect tenants from unscrupulous landlords. However, many politicians argued that rent controls distorted the housing market and they are no longer popular.

repo rate The interest rate at which the Bank of England lends on a short-term basis to the UK banking sector.

research and development policy Research and development (R&D) is undertaken by a wide range of institutions and organisations, although about 50 per cent of UK R&D is government funded. Historically, the UK has been strong in research but less effective in development.

resource allocation The assignment of resources to various uses. More specifically, it means determining what will be produced, how it will be produced, who will produce it, and for whom it will be produced.

resources Inputs used in the production of goods and services. Commonly separated into four categories: land, labour, capital and entrepreneur (see separate glossary entries for details). Also called factors of production.

retail price index (RPI) A statistical measure of the change in prices of a representative set of goods and services purchased by the average household in the UK.

right to buy The Housing Act 1980 gave council tenants in the UK the right to acquire their houses or flats at a discount of the market value and effectively allowed local authorities to transfer their housing stock to the private sector.

Royal Institute of British Architects (RIBA) The principal professional body in the United Kingdom concerned with architecture. Established in 1834, it currently has 27,000 members, with 6,000 registered overseas.

Royal Institution of Chartered Surveyors (RICS) The main UK professional body concerned with surveying in its various guises. It was founded in 1868 and now has over 85,000 members across its seven different divisions.

scarcity A reference to the fact that at any point in time there is a finite amount of resources, while people have an infinite amount of 'wants' for goods and services.

services Goods that do not have physical characteristics. Examples of services include the 'goods' provided by doctors, lawyers, dentists, educators, retailers, surveyors, wholesalers and welfare staff.

short run The time period in which a firm cannot alter its current size of plant.

sick building syndrome Defined by the World Health Organisation as a syndrome of complaints covering general feelings of malaise, the onset of which is associated with the occupancy of certain modern buildings. The problem particularly affects the health of some office workers.

social costs The full cost that society bears when a resource-using action occurs. For example, the social cost of driving a car is equal to all the private costs plus any additional cost that society bears, such as air pollution and traffic congestion. (Some authors use this term to simply refer to external costs.)

social price The total price when *all* costs and benefits have been considered – that is, when private costs and benefits are added to the external costs and benefits.

stable equilibrium A situation in which if there is a shock that disturbs the existing relationship between the forces of supply and demand, there are self-corrective forces that automatically cause the disequilibrium to be remedied.

stagflation A term coined in the 1970s to represent the combined problem of high unemployment and rising prices.

strategic partnering An agreement between a contractor and client to work together on a series of construction projects in order to promote continuous improvement. Contrast with project partnering.

structural rigidities These are obstacles within markets that prevent a swift response to changing forces of supply and demand. They are more prevalent in some markets than others. One example of a structural rigidity is the way that commercial leases affect the dynamics of the property market.

subnormal profits A rate of return that is below the rates being earned elsewhere. More commonly these would be referred to as a loss.

substitute goods Two goods are considered substitutes if one can be used in place of the other. A change in the price of one, therefore, causes a shift in demand for the other. For example, if the price of butter goes up, the demand for margarine will rise; if the price of butter goes down, the demand for margarine will decrease.

supernormal profits A rate of return that is greater than the rates being earned elsewhere. Also known as abnormal profits.

supply constraints These occur when it is either not possible to increase output – or, it is only possible at high costs.

supply curve The graphical representation of the supply schedule; a line showing the supply schedule, which slopes upwards (has a positive slope).

supply schedule A set of numbers showing prices and the quantity supplied at various prices for a specified period of time. In effect, therefore, the schedule shows the rate of planned production at each price.

supply-side economics Government policies designed to create incentives for firms and individuals to increase productivity. Supply-side economics is concerned with the level of aggregate supply.

sustainable construction An international agenda set for the construction industry designed to deliver products that perform better over their full life cycle and work in a more economically efficient and socially responsible way. For other definitions, see Table 15.4 (page 274).

sustainable development In general terms, developments that balance social, environmental and economic concerns. The sustainable development agenda recognises that decisions made today can have serious implications for future generations. For other definitions see Table 15.1 (page 269), which includes the most widely quoted definition from the World Commission on Environment and Development.

sustainable investment rule A fiscal rule stating that public sector net debt as a proportion of GDP should be held over the economic cycle at a stable level – that is, below 40 per cent of GDP.

tax bracket A band of income to which a specific and unique marginal tax rate is applied.

tax burden The incidence of tax within society as a whole – and the contribution in tax of different sections of society.

theory of the firm A theory of how suppliers of any product or service make choices in the face of changing constraints.

third party Persons external to negotiations and activities between buyers and sellers. If person A buys a car with no brakes and then runs person B over, person B is a third party to the deal struck between person A and the seller of the car, and person B's suffering is a negative externality.

tightly coupled systems An industry typified by close relationships and high levels of co-ordination between producers. For example, manufacturing displays many tightly coupled systems. Contrast with loosely coupled systems.

total costs All the costs of a firm combined, such as rent, payments to workers, interest on borrowed money, rates, material costs, etc.

total expenditure The total monetary value of all the final goods and services bought in an economy during a year.

total income The total amount earned by the nation's resources (factors). National income, therefore, includes wages, rent, interest payments and profits that are received, respectively, by workers, landowners, capital owners and entrepreneurs.

total output The total value of all the final goods and services produced in the economy during the year.

total revenue The price per unit times the total quantity sold.

trade-off A term relating to opportunity cost. In order to get a desired economic good, it is necessary to trade off some other desired economic good whenever we are in a world of scarcity. A trade-off involves the sacrifice that must be made in order to obtain something.

transaction costs All the costs associated with exchanging, such as the informational costs of finding out price and quality, service record and durability, and the cost of enforcing the contract.

transfer payments Money payments made by governments to individuals for which no services or goods are concurrently rendered. Examples are social security payments and student grants.

transition economies Refers to the economies of the 25 countries in Eastern Europe and the former Soviet Union that are moving from a centrally planned to a market system. The term, however, can be used less strictly to refer to any economy changing its system of resource allocation.

travel cost method A technique used to identify the price of an externality. The central premise involves estimating how much people are willing to pay to travel in order to experience an environmental asset.

unit elastic demand If a percentage change in price leads to an identical percentage change in demand, then the product has an elasticity of unity – that is, an elasticity of one.

unit elastic supply A property of the supply curve where the quantity supplied changes exactly in proportion to changes in price. In this hypothetical situation, revenue is invariant to price changes.

u-value A traditional measurement of heat loss. As the u-value coefficient moves nearer to zero the insulation quality of the material being measured improves. In other words, as the u-value of a wall lowers there is less heat lost through the fabric of the building.

value management A term used to embrace all activities and techniques used to establish where costs are being incurred and where value could be added. In effect, value management is a process that should achieve better value for money for the client. It is commonly abbreviated to VM.

variable costs Costs that vary with the rate of production. Examples include wages paid to workers, the costs of materials, and so on.

vendor A seller – especially one who sells land and property.

vertical equity The concept behind measures to achieve social justice or fairness by providing benefits targeted at people with specific needs. The idea that underpins policies such as means-tested benefits and taxing the rich more heavily than the poor.

voluntary instruments This term has emerged as the green agenda has developed, and it refers to the various assessment schemes for buildings and management. Ironically, the common thread is that these schemes always have to be 'paid for' as legislation has not been introduced to make them compulsory.

wage-price spiral An inflationary process in which incomes and retail prices follow each other in an upward direction (see Figure 14.3).

wages councils Bodies set up by the government to determine the pay of those in occupations that are traditionally poorly rewarded. At their peak these bodies regulated wages for more than one million employees in trades such as retailing and agriculture. Wages councils were abolished in the UK in 1993.

x-inefficiency Describes the organisational slack that results in higher unit costs than would occur within a more competitive marketplace.

References

Ball, M., Farshchi, M. and Grilli, M. (2000) 'Competition and the Persistence of Profits in the UK Construction Industry', *Construction Management and Economics*, 18: 733–45

Balls, E. and O'Donnell, G. (eds) (2002) *Reforming Britain's Economic and Financial Policy: Towards greater economic stability*, Palgrave: Basingstoke

Banwell, H. (1964) *The Placing and Management of Contracts for Building and Civil Engineering Works*, HMSO: London

Barbier, E.B. (1987) 'The concept of sustainable economic development', *Environmental Conservation*, 2: 101–11

Barker, K. (2003) *Review of Housing Supply: securing our future needs – Interim report*, HM Treasury: London

Barwell, R. Thomas, R. and Turnbull, K. (2007) 'The macroeconomic impact of higher energy prices on the UK economy', *Bank of England Quarterly Bulletin*, Winter, Bank of England: London

Baumol, W.J. (1982) 'Contestable markets: an uprising in the theory of industrial structure', *American Economic Review*, 72: 1–15

Bennett, J. and Jayes, S. (1995) *Trusting the Team*, University of Reading: Centre for Strategic Studies in Construction

BERR – Department for Business Enterprise and Regulatory Reform (2007) *Construction Statistics Annual*, TSO: London

BERR – Department for Business Enterprise and Regulatory Reform (2008) *Construction Output Time Series Data*, available in Excel format from www.berr.gov.uk

Bill, T. (2008) 'Construction firms will take credit crunch hit in 2008', *Building*, 11 January

Blanchflower, D. (2008) *Inflation, expectations and Monetary Policy*, speech delivered at the Royal Society, 29 April. Available from Bank of England: London

Blismas, N., Pasquire, C. and Gibb, A. (2006) 'Benefit evaluation for off-site production in construction', *Construction Management and Economics* 24: 121–30

Bon, R. (1992) 'The future of international construction: secular patterns of growth and decline', *Habitat International*, 16(3), 119–28

Bon, R. and Crosthwaite, D. (2000) *The Future of International Construction*, Thomas Telford: London

Boulding, K.E. (1966) 'The economics of the coming spaceship earth', in H. Jarrett (ed.) *Environmental Quality in a Growing Economy*, John Hopkins University Press: Baltimore

Broughton, T. (2001) 'In Person: Brian Wilson Interview', *Building*, 14 December: 26–7

Building (2008) 'Egan: house builders have done absolutely nothing to improve', *Building,* 23 May: 10

Carassus, J. (ed.) (2004) *The construction sector system approach: an international framework*, CIB paper 293, International Council for Research and Innovation in Building and Construction: Rotterdam.

Chan, P.C., Chan, D.W.M. and Ho, K.S.K. (2003) 'An empirical study of the benefits of construction partnering in Hong Kong', *Construction Management and Economics,* 21: 523–33

Chan, S. (2002) 'Responses of selected economic indicators to construction output shocks: the case of Singapore', *Construction Management and Economics,* 20: 523–33

Chatterji, M. and Tilley, C. (2002) 'Sickness, absenteeism, presenteeism, and sick pay', *Oxford Economic Papers,* 54: 669–87

Chen, J.J. (1998) 'The characteristics and current status of China's construction industry', *Construction Management and Economics,* 16: 711–19

Chevin, D. (1992) 'Wimpey shelves green homes', *Building,* 24 April: 7

Chevin, D. (2008) 'China goes to town', *Building,* 4 July: editorial

Chiang, Y., Tang, B. and Wang, F.K.W. (2008) 'Volume building as competitive strategy', *Construction Management and Economics,* 26: 161–76

Cooke, A.J. (1996) *Economics and Construction,* Macmillan: London

Costanza, R., D'Arge, R., De Groot, R., Farber, S., Grasso, M., Hannon, B., Limburg, K., O'Neill, R., Parnefo, J., Raskin, R. and Sutton, P. (1997) 'The value of the world's ecosystem services and natural capital', *Nature,* 387: 253–60

Cox, R. and Goodman, C. (1956) 'Marketing of house building materials', *Journal of Marketing,* 21: 36–61

Crosthwaite, D. and Connaughton, J. (2008) *World Construction 2007–2008,* Davis Langdon Management Consulting: London

Daly, H.E. (1999) 'Uneconomic growth and the built environment: In theory and in fact', in C.K. Kibert (ed.) *Reshaping the Built Environment: Ecology, Ethics and Economics,* Island Press: Washington

Davis, Langdon and Everest (2000) *European Construction Costs Handbook,* Spon: London

DCLG – Department for Communities and Local Government (2008) *Greener homes for the future,* DCLG: London

DCLG – Department for Communities and Local Government (2007) *The Future for Building Control,* DCLG: London

De Valence, G. (2006) Guest Editorial, *Construction Management and Economics,* 24: 661–8

Defra – Department for Environment, Food and Rural Affairs (2005) *UK Strategy for Sustainable Development* Defra: London

Defra – Department for Environment, Food and Rural Affairs (2007) *An introductory guide to valuing ecosystem services,* Defra: London

Deloitte (2006) *Construction Demand/ Capacity Study – 2000 to 2015,* Office of Government and Commerce: London

Desai, P. and Riddlestone, S. (2002) *Bioregional Solutions: For living on one planet,* Green Books: Devon

DETR – Department of the Environment, Transport and the Regions (2000) *Building a Better Quality of Life: A Strategy for More Sustainable Construction,* DETR: London

Dey-Chowdhury, S. (2007) 'House price indices of the UK', E*conomic and Labour Market Review,* 1: 54–8

DfT – Department for Transport (2007) *Adding Capacity At Heathrow Airport: consultation document,* DfT: London

DLC – Davis Langdon Consultancy (1998) *Review of Construction Markets in Europe,* April, Davis Langdon Consultancy: London

Drew, D. and Skitmore, P. (1997) 'The effect of contract type and size on competitiveness in bidding', *Construction Management and Economics,* 15: 469–89

Du Plessis, C. (2002) *Agenda for Sustainable Construction in Developing Countries: A discussion document,* report for CIB and UNEP available at www.cibworld.nl

Du Plessis, C. (2005) 'Action for sustainability: preparing an African plan for sustainable building and construction', *Building Research and Information,* 33: 405–15

Dubios, A. and Gadde, L. (2002), 'The construction industry as a loosely coupled system: implications for productivity and innovation', *Construction Management and Economics,* 20: 621–31

Dulaimi, M.S., Ling, F.Y.Y. and Ofori, G. (2001) *Building A World Class Construction Industry: Motivators and Enablers,* Singapore University Press: Singapore

Economist, The (2002) 'Gordon Brown's Manifesto', *The Economist,* 20 April

EEA – European Environment Agency (2008) *Effectiveness of environmental taxes and charges for managing sand, gravel and rock extraction in selected EU countries,* European Environment Agency: Copenhagen

Egan, J. (1998) Rethinking Construction: *The report of the construction task force,* DETR: London

Egan, J. (2002) *Rethinking Construction: Accelerating change,* Strategic Forum Secretariat: London

Elkins, P. (2008)'Path of least resistance', *The Guardian,* 13 February

Elliott, J. (1999) *An introduction to sustainable development,* Routledge: London and New York

Ellis, R.C.T., Wood, G.D. and Keel, D.A. (2005) 'Value management practices of leading UK cost consultants', *Construction Management and Economics,* 23: 483–93

Environment Agency (2007) S*potlight on business environmental performance in 2006,* Environment Agency: England

EU – European Union (2001) *Competitiveness of the Construction Industry: An agenda for sustainable construction in Europe,* European Commission: Brussels

EU – European Union (2007) *Accelerating the Development of the Sustainable Construction Market in Europe,* European Commission: Brussels

Evans, R., Haryott, R., Haste, N. and Jones, A. (1998) *The long-term costs of owning and using buildings*, Royal Academy of Engineering: London

Fairclough, J. (2002) *Rethinking Construction Innovation and Research: A review of government R&D policies and practices*

Fellner, W. (1949) *Competition Among the Few: Oligopoly and Similar Market Structures*, Alfred A. Knoff: New York

FIEC – European Construction Industry Federation (2007) *Construction in Europe Key Figures*, Fédération de l'Industrie Européenne de la Construction: Brussels

Flyvbjerg, B. (2003) *Megaprojects and Risk: an anatomy of ambition*, Cambridge University Press: Cambridge

Forum for The Future (2002) *Sustainability Accounting in the Construction Industry*, CIRIA: London

Friedman, M. (1962) *Capitalism and Freedom*, (with new preface in the 40th anniversary edition in 2002), University of Chicago Press: Chicago and London

Gann, D. (2000) *Building Innovation: Complex Constructs in a Changing World*, Thomas Telford: London

GFN – Global Footprint Network (2006) *Ecological Footprint and Biocapacity: technical notes*, data and methods available from www.footprintnetwork.org

Gidado, K.I. (1996) 'Project complexity: the focal point of construction production planning', *Construction Management and Economics*, 14: 213–25

Goodland, R., Daly, H. and El Serafy, S. (1992) *Population, technology and lifestyle*, Island Press: Washington DC

Gruneberg, S. (1997) *Construction Economics: An Introduction*, Macmillan: London

Gruneberg, S (2008) 'Why some firms count on bid rigging to survive', www.building.co.uk/oftinvestigation [accessed 22 April 2008]

Gruneberg, S. and Ive, G. (2000) *The Economics of the Modern Construction Firm*, Macmillan: London

Gummer, J. and Goldsmith, Z. (2007) *Blueprint for a Green Economy: quality of life policy group, submission to the shadow cabinet*, Conservative Party: London,

Gunny, S. (2005) 'Stay Free', *Building*, 24 March

Gwilliam, K. (2000) 'The new economics of sustainable transport in developing countries: incentives and institutions', in J.B. Polak and A. Heertje (eds) *Analytical Transport Economics: an international perspective*, Edward Elgar: Cheltenham, UK and Northampton MA, USA

Han, S.S. and Ofori, G. (2001) 'Construction industry in China's regional economy, 1990–1998', *Construction Management and Economics*, 19: 189–205

Hedges, A. and Wilson, S. (1987) *The office environment survey: A study of building sickness*, Building Use Studies: London

Heerwagen, J. (2000) 'Green buildings, organizational success and occupant productivity', *Building Research and Information*, 28: 353–67

Hellowell, M. and Pollock, A.M. (2007) *Private Finance, Public Deficits: A report on the cost of PFI and its impact on health services in England*, Centre for International Public Health Policy, University of Edinburgh: Scotland

Hillebrandt, P. (1974) *Economic Theory and the Construction Industry,* Macmillan: London

Hillebrandt, P. (1984) *Analysis of the British Construction Industry,* Macmillan: London

Hillebrandt, P. (1985) *Economic Theory and the Construction Industry,* second edition, Macmillan: London

Hillebrandt, P. (2000) *Economic Theory and the Construction Industry,* third edition, Macmillan: London

HM Government (2008) *Strategy for Sustainable Construction,* BERR: London

HM Treasury (1997) *Statement of Intent on Environmental Taxation,* HMSO: London

HM Treasury (2002) *Tax and the Environment Using Economic Instruments,* TSO: London

HM Treasury (2006) *Value for Money Assessment Guidance,* HMSO: London

HM Treasury (2007), *PFI Signed Project List – October 2007,* available as an Excel file from www.hm-treasury.gov.uk

HM Treasury (2008), *Forecasts for the UK economy: A comparison of independent forecasts – May 2008,* No. 253, monthly update available from www.hm-treasury.gov.uk/forecasts

ICLEI – International Council for Local Environmental Initiatives (1996) *The Local Agenda 21 Planning Guide*

International Monetary Fund (1998) *Code of Good Practices on Fiscal Transparency,* IMF: Washington DC

International Monetary Fund (2007) *World Economic Outlook: Globalization and Inequality,* IMF: Washington DC

IPCC – Intergovernmental Panel on Climate Change (2007) *Climate Change 2007: the Fourth IPCC Assessment Report,* Cambridge University Press: New York and Cambridge

Ive, G. and Gruneberg, S. (2000) *The Economics of the Modern Construction Sector,* Macmillan: London

Jensen, J.O. and Gram-Hansenn, K. (2008) 'Ecological modernization of sustainable buildings: a Danish perspective', *Building Research and Information,* 36: 146–58

Jessop, S. (2002) 'Environmental Supply Chain Management: A Contractors Perspective', in CIEF meeting notes *Environment and the Supply Chain – the weakest link?,* 16 April, London

Kaan, H.F. and de Boer, B.J. (2006) *Passive houses: achievable concepts for low CO_2 housing,* report ECN-RX–06-019, Energy Research Center: Netherlands

Kibert, C. (1994) 'Principles of sustainable construction', in *Proceedings of the 1st International Conference on Sustainable Construction,* Tampa, Florida, US, 6–9 November: 1–9

Kibert, C.J. (ed.) (1999) *Reshaping the Built Environment: Ecology, Ethics and Economics,* Island Press: Washington DC

Kibert, C. (2007) 'The next generation of sustainable construction', *Building Research and Information,* 35: 595–601

Latham, M. (1994) *Constructing the Team,* HMSO: London

Leibenstein, H. (1973) 'Competition and X-Inefficiency: Reply', *Journal of Political Economy,* 81: 765–77

Lipsey, R. and Crystal, A. (1995) *An Introduction to Positive Economics,* Oxford University Press: Oxford

Lovins, A., Lovins, H. and Hawken, P. (1999) 'A road map for natural capitalism', *Harvard Business Review,* May-June: 145–58

Lowe, J. (1987) 'Monopoly and the materials supply industres of the UK', *Construction Management and Economics,* 5: 57–71

Marx, K. (1844) *Economic and Philosophical Manuscripts,* Pelican Marx Library: London

Manseau, A. and Seaden, G. (eds) (2001) *Innovation in Construction: An International Review of Public Policies,* Spon Press: London and New York. A paper presenting the framework and conclusions preceded the comprehensive detailed report in *Building Research and Information,* 2001, 29: 182–96

Matar, M.M., Georgy, M.E. and Ibrahim, M.E. (2008) 'Sustainable construction management: introduction of the operational context space (OCS)', *Construction Management and Economics,* 26: 261–75

McMeeken, R. (2008) 'The Squeeze', *Building,* 15 February: 41–3

Meadows, D.H., Meadows, D.L., and Randers, J. (1992) *Beyond the Limits – Global Collapse or a Sustainable Future,* Earthscan Publications: London

Meadows, D.H., Meadows, D.L., Randers, J., and Behrens, W. (1972) *The Limits to Growth – A Report to the Club of Rome,* Universe Books: New York`

Meadows, D.H., Richardson, J., and Bruckmann, G. (1982) *Groping in the Dark – The First Decade of Global Modelling,* John Wiley: New York

Mehmet, O. and Yorucu, V. (2008) 'Explosive construction in a micro-state: environmental limit and the Bon curve: evidence from North Cyprus', *Construction Management and Economics,* 26: 79–88

Miller, T.R. (2000) 'Variations between countries in values of statistical life', *Journal of Transport Economics and Policy,* 34: 169–88

MPC – Monetary Policy Committee (1999) 'The transmission mechanism of monetary policy' *Bank of England Quarterly Bulletin,* Spring, Bank of England: London

Myers, D. (2005) 'A review of construction companies' attitudes to sustainability', *Construction Management and Economics,* 23: 781–85

NAO – National Audit Office (2001) *Modernising Construction,* TSO: London

NAO – National Audit Office (2005) *Using Modern Methods of Construction to Build Homes More Quickly and Efficiently,* TSO: London

NAO – National Audit Office (2007) *Homebuilding: Measuring Construction Performance,* TSO: London

NAO – National Audit Office (2007a) *The Carbon Trust: accelerating the move to a low carbon economy,* TSO: London

NAO – National Audit Office (2008) *Making Changes in Operational PFI Projects,* TSO: London

NCBS – National Centre for Business and Sustainability (2003) *Rock to Rubble: building a sustainable region,* A 4sight Report, UK: Manchester

Ofori, G. (1994) 'Establishing Construction Economics as an Academic Discipline', *Construction Management and Economics*, 12: 295–306

Ofori, G. (2002) 'Singapore's construction: moving toward a knowledge-based industry', *Building Research and Information*, 30: 401–12

OGC – Office of Government Commerce (2006) *Making competition work for you: A guide for public sector procurers of construction*, OGC: London

OGC – Office of Government Commerce (2007) *Value management in construction: Case Studies*, OGC: London

ONS – Office for National Statistics (2000) *Guide to Official Statistics*, ONS: London

ONS – Office for National Statistics (2006) *UK Input-output analyses*, ONS: London

ONS – Office for National Statistics (2007) *UK National Accounts: The Blue Book*, ONS: London

Osborne, H. (2001) 'Growing eco-assets', *Estates Gazette*, 25 August: 84–5

Pan, W., Gibb, A.G.F. and Dainty, A.R.J. (2008) 'Leading UK housebuilders' utilization of offsite construction methods', *Building Research and Information*, 36:1, 56–67

Pearce, D. (2003) *The Social and Economic Value of Construction: the construction industry's contribution to sustainable development*, nCRISP: London

Pietroforte, R., Costantino, N., Falagario, M. and Gould, J.S. (2007) A review of the abstracts of construction management and economics, 1983–2006, *CME Conference Proceedings*, University of Reading

Pezzey, J. (1989) *Definitions of sustainability*, UK CEED: Peterborough

Pratten, C.F. (1971) *Economies of Scale in Manufacturing Industry*, Cambridge University Press: Cambridge

Reeves, K. (2002) 'Construction business systems in Japan: general contractors and subcontractors', *Building Research and Information*, 30: 413–24

Ricardo, D. (1817) *On the Principles of Political Economy and Taxation*, John Murray: London

Richardson, S. (2008) 'Construction industry is on brink of recession', *Building*, 11 January: 10

Ruddock, L. and Lopes, J. (2006) 'The construction sector and economic development: the Bon curve', *Construction Management and Economics* 24: 717–23

Samuelson, P.A. and Nordhaus, W.D. (2005) *Economics International Edition*, 18th edition, McGraw Hill: New York

Sayce, S., Ellison, L. and Parnell, P. (2007) 'Understanding investment drivers for UK sustainable property', *Building Research and Information*, 35: 629–43

Schot, J., Brand, E. and Fischer K. (1997) *The Greening of Industry for a Sustainable Future: Building an International Research Agenda*, The Greening of Industry Network: The Netherlands

Shash, A.A. (1993) 'Factors considered in tendering decisions by top UK contractors', *Construction Management and Economics*, 11: 111–18

Shiers, D. (2000) 'Environmentally responsible buildings in the UK commercial property sector', *Property Management*, 18: 352–65

Smith, R.A., Kersey, J.R. and Griffiths, P.J. (2003) *The Construction Industry Mass Balance: resource use, wastes and emissions*, Viridis Report VR4, Viridis and CIRIA: London

Solow, R. (1974) 'The economics of resources or the resources of economics', *Amercian Economic Review*, 64: 1–14

Stern, N. (2007) *The economics of climate change: the Stern review*, Cambridge University Press: Cambridge. A web edition was released in December 2006, it is available as an independent review from www.hm-treasury.gov.uk

Sustainable Construction Task Group (2002) *Reputation, Risk and Reward: the business case for sustainability in the UK property sector*, BRE: Watford

Sweet, R. (2002) 'The Long March', *Construction Manager*, April: 17–19

Turin, D.A. (1967) *Building as a process, proceedings of the Bartlett Society*: London. Reprinted in Building Research and Information, 2003, 31: 180–7

Turin, D.A. (ed.) (1975) *Aspects of the Economics of Construction*, George Godwin: London

UK Trade and Investment (2007) *Transportation: India's Sunrise Sector*, executive summary available from www.uktradeinvest.gov.uk

United Nations (2005) *Economic Survey of Europe, 2005 No. 1*, United Nations: New York and Geneva

Urge-Vorsatz, L.D., Harvey, D., Sevastianos, M. and Levine, M.D. (2007) 'Mitigating CO_2 emissions from energy use in the world's buildings', *Building Research and Information*, 35: 379–98

WCED – World Commission on Environment and Development (1987) *Our Common Future*, Oxford University Press: Oxford

Weizsäcker, E. von, Lovins, A. and Lovins, H. (1998) *Factor Four: Doubling Wealth, Halving Resource Use*, Earthscan: London

Winch, G. (1987) 'The construction firm and the construction process: the allocation of resources to the construction project', in P. Lansley and P. Harlow (eds), *Managing Construction Worldwide, Volume 2, Productivity and Human Factors*, E & FN Spon: London

World Bank (2008) *World Development Indicators*, World Bank: Washington DC

WWF – World Wide Fund for Nature (2006) *Living Planet Report*, WWF International: Switzerland.

Xiao, H. and Proverbs, D. (2002) 'The Performance of Contractors in Japan, the UK, and the USA: a comparative evaluation of construction cost', *Construction Management and Economics*, 20: 425–35

Yang, W. and Kohler, N. (2008) 'Simulation of the evolution of the Chinese building and infrastructure stock', *Building Research and Information*, 36: 1–19

Index